W9-CDX-342

Global Warming and Climate Change

Current Issues

San Diego, CA

Other books in the Compact Research series include:

Drugs

Alcohol
Club Drugs
Cocaine and Crack
Hallucinogens
Heroin
Inhalants
Marijuana
Methamphetamine
Nicotine and Tobacco
Performance-Enhancing Drugs

Current Issues

Biomedical Ethics
The Death Penalty
Energy Alternatives
Free Speech
Gun Control
Illegal Immigration
National Security
Nuclear Weapons and Security
Terrorist Attacks
World Energy Crisis

Global Warming and Climate Change

by Emma Carlson Berne

Current Issues

San Diego, CA

For more information, contact
ReferencePoint Press, Inc.
PO Box 27779
San Diego, CA 92198
www. ReferencePointPress.com

Picture credits:
AP/Wide World Photos, 14, 15
Maury Aaseng, 36–39, 53–56, 68–71, 85–88
Photos.com, 13

Series design:
Tamia Dowlatabadi

LIBRARY OF CONGRESS CATALOGING-IN-PUBLICATION DATA

Berne, Emma Carlson
Global warming and climate change / by Emma Carlson Berne.
p. cm. — (Compact research series)
Includes bibliographical references and index.
ISBN-13: 978-1-60152-019-7 (hardback)
ISBN-10: 1-60152-019-0 (hardback)
1. Global warming. 2. Climatic changes. I. Title.
QC981.8.G56B454 2006
363.738'74—dc22

2007008371

Contents

Foreword

“Where is the knowledge we have lost in information?”

—“The Rock,” T.S. Eliot.

As modern civilization continues to evolve, its ability to create, store, distribute, and access information expands exponentially. The explosion of information from all media continues to increase at a phenomenal rate. By 2020 some experts predict the worldwide information base will double every 73 days. While access to diverse sources of information and perspectives is paramount to any democratic society, information alone cannot help people gain knowledge and understanding. Information must be organized and presented clearly and succinctly in order to be understood. The challenge in the digital age becomes not the creation of information, but how best to sort, organize, enhance, and present information.

ReferencePoint Press developed the Compact Research series with this challenge of the information age in mind. More than any other subject area today, researching current events can yield vast, diverse, and unqualified information that can be intimidating and overwhelming for even the most advanced and motivated researcher. The Compact Research series offers a compact, relevant, intelligent, and conveniently organized collection of information covering a variety of current and controversial topics ranging from illegal immigration to marijuana.

The series focuses on three types of information: objective single-author narratives, opinion-based primary source quotations, and facts and statistics. The clearly written objective narratives provide context and reliable background information. Primary source quotes are carefully selected and cited, exposing the reader to differing points of view. And facts and statistics sections aid the reader in evaluating perspectives. Presenting these key types of information creates a richer, more balanced learning experience.

For better understanding and convenience, the series enhances information by organizing it into narrower topics and adding design features that make it easy for a reader to identify desired content. For example, in *Compact Research: Illegal Immigration*, a chapter covering the economic impact of illegal immigration has an objective narrative explaining the various ways the economy is impacted, a balanced section of numerous primary source quotes on the topic, followed by facts and full-color illustrations to encourage evaluation of contrasting perspectives.

The ancient Roman philosopher Lucius Annaeus Seneca wrote, "It is quality rather than quantity that matters." More than just a collection of content, the Compact Research series is simply committed to creating, finding, organizing, and presenting the most relevant and appropriate amount of information on a current topic in a user-friendly style that invites, intrigues, and fosters understanding.

Global Warming and Climate Change at a Glance

Rising Temperatures

Earth is growing warmer. As of 2006 Earth's temperatures were the warmest they had been in the last 42,000 years.

Greenhouse Gases

Most scientists attribute the rise in temperature to an increase in "greenhouse gases" present in Earth's atmosphere. These are gases that thicken Earth's atmosphere, trapping the sun's radiant heat inside.

Emissions

Most greenhouse gases are added to the atmosphere by the burning of coal, natural gas, and other fossil fuels. There are 30 types, but 80 percent of all greenhouse gases in the atmosphere is carbon dioxide.

Rapid Growth

Temperatures have risen in proportion to greenhouse gas levels since the advent of industrialization. Carbon dioxide levels are expected to hit double their preindustrial levels sometime in the mid–twenty-first century.

Consequences

Warmer temperatures will have profound social, physical, and economic consequences. Melting ice, increased precipitation, and differences in ocean currents change climates, food supplies, and ecological systems.

Scientific Views vs. Public Opinion

Scientists and the public have long differed on whether global warming is actually happening. The gap may be closing, though. A 2006 poll showed that the majority of the American public believes that Earth is getting warmer.

Global Efforts

The global community has taken some steps to limit greenhouse gas emissions. The Kyoto Protocol sets limits on the amount of carbon dioxide the world's industrialized countries can produce each year.

State Efforts

After scant attention to global warming by the federal government, many states have initiated their own carbon reduction programs. California has led the movement, requiring a 25 percent cut in emissions by 2020.

Overview

“Human activities have become a major source of environmental change.”

—Consensus statement of the American Meteorological Association, February 2003.

“Climate has varied on every time scale to which we have any observational access.”

—Richard J. Somerville, “What’s Up With the Weather?” April 2000.

What Is Global Warming and Climate Change?

The atmosphere that surrounds Earth is composed of a thin layer of naturally occurring gases. This crucial layer protects the planet from the heat of the sun and the cold of space and keeps Earth at a life-sustaining temperature through a process called the greenhouse effect. Radiant heat and light from the sun passes through the thin atmosphere to the planet itself. Some of this energy warms the ground and waters of Earth and then reradiates back into space through the atmosphere. Some of the sun’s energy, though, gets trapped by the atmosphere. This heat and light does not reradiate into space. Instead, it warms Earth’s air, which would otherwise be a frigid zero degrees Fahrenheit; the greenhouse effect is necessary to sustain life as we know it on Earth.

Global warming occurs when certain gases—called greenhouse gases—thicken the thin layer of atmosphere, trap too much heat inside, and further warm the ground, oceans, air, and clouds. The primary greenhouse gas is carbon dioxide, but there are about 30 others. Some of these are present in the air, land, and oceans naturally, others have been created by humans.

The planet cycles through natural cooling and warming periods on its own. These can last for a few hundred years or a few millennia. Significant, sustained changes in temperature such that plant and animal life are affected is called "climate change." For the last half-century, climate change has been occurring in the form of a warming trend. Some scientists believe that this particular change is not natural, however. They have concluded that the presence in the atmosphere of greenhouse gases from human activity, especially carbon dioxide, have contributed to the warming that Earth is currently experiencing.

> **"Global warming occurs when certain gases—called greenhouse gases—thicken the thin layer of atmosphere, trap too much heat inside, and further warm the ground, oceans, air, and clouds."**

Energy Generation

The elements that make modern life possible around the world—electrical plants that run on fossil fuels, transportation by combustion engine, large-scale agriculture—also emit greenhouse gases into the atmosphere. As the global population increases, and energy needs increase along with it, the demand for cheap coal-based energy, natural gas, and petroleum also grows. Power generation, for instance, produces almost 25 percent of all carbon dioxide emissions worldwide. Seventy percent of American electricity needs are generated by the burning of fossil fuels (coal, petroleum, and natural gas), major emitters of carbon dioxide. Fifty percent of all Americans get their electricity from a particularly dirty fossil fuel: coal, with energy generated from coal-fired power plants. These power plants—in the United States and elsewhere in the world—are one of the primary contributors to the global warming problem.

Other sectors also contribute to climate change. Deforestation, for example, because it removes trees that take in carbon dioxide and release oxygen, contributes 18 percent of the total amount of global greenhouse gas emissions. Gasoline-based transportation—cars, trucks, airplanes, ships—contributes about 13 percent.

The Current and Future Growth of Greenhouse Gases

Carbon dioxide emissions have increased since measurements began in the 1950s. Many scientists attribute this to the rise of industrialization that began after World War II. Though no official measurements exist, scientists have estimated that current carbon dioxide levels are one and a half times higher than they were during preindustrialization. If growth continues unchecked, carbon dioxide levels will rise to double their preindustrial levels by the middle of the twenty-first century.

In her book *Field Notes from a Catastrophe*, science writer Elizabeth Kolbert notes that to avoid hitting double preindustrial levels, global carbon dioxide emissions would have to be held at zero immediately. In addition, carbon dioxide exists in the atmosphere for approximately a century, so even if the global population were to immediately halt all carbon production, carbon levels would continue to rise.

> **"Increased greenhouse gas levels mean a warmer Earth."**

Increased greenhouse gas levels mean a warmer Earth. Jim Hansen, director of the Goddard Institute for Space Studies, wrote in a recent paper that Earth is only one degree Celsius from its hottest level in 1 million years. This represents a sort of threshold for climate scientists with regard to global warming: Many predict that after that point, the effects of global warming will be impossible or nearly impossible to reverse. Many estimate that one more decade of emissions growth at the current rate will increase global warming to the point that this increase will be impossible to avoid.

What Are the Consequences of Global Warming?

The consequences of global warming and climate change are a result of the primary effect: higher temperatures on Earth. Secondary consequences include rising sea levels, warming of permafrost, increased natural disasters, and increased presence of disease, as well as strain on local and global economies. Previously existing problems familiar to humans since the beginning of their existence can become exacerbated by global warming. One of the more significant of these is the increase in the intensity

Deforestation is detrimental to the environment because it removes trees that take in carbon dioxide and release oxygen. Deforestation contributes 18 percent of the total amount of global greenhouse gas emissions.

of hurricanes that increased global temperatures will cause, although this theory is still under debate within the scientific community.

Increase in Sea Levels and Flooding

As global warming continues, the world's glaciers are melting, including those in the European Alps, the Himalayas, and the territory of Greenland. Snow cover is also declining. This melt is having an effect on the world's oceans. As the massive ice sheets diminish, fresh water pours into the sea and raises sea levels. Even a .04-inch or .08-inch rise (1mm or 2mm) can increase flooding near coastlines. The Worldwatch Institute, a nonprofit environmental group, reported in 2006 that sea levels have risen 394 feet (120m) since the peak of the last Ice Age 21,000 years ago.

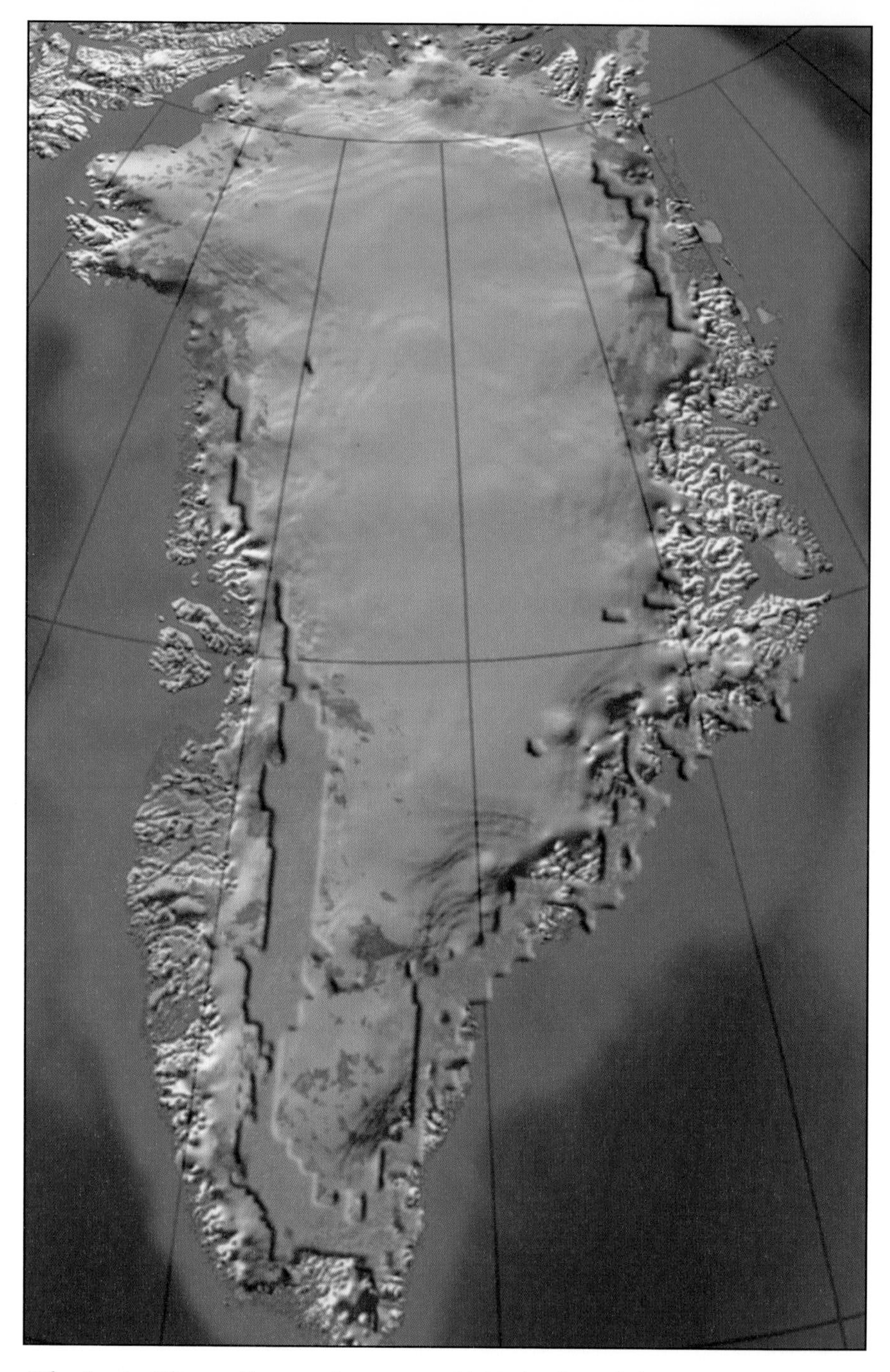

The Arctic Climate Impact Association (ACIA) released this image of the Greenland Ice Sheet in 1992. An image of the same area released by the group in 2002 (opposite) shows how much of the sheet melted in just 10 years.

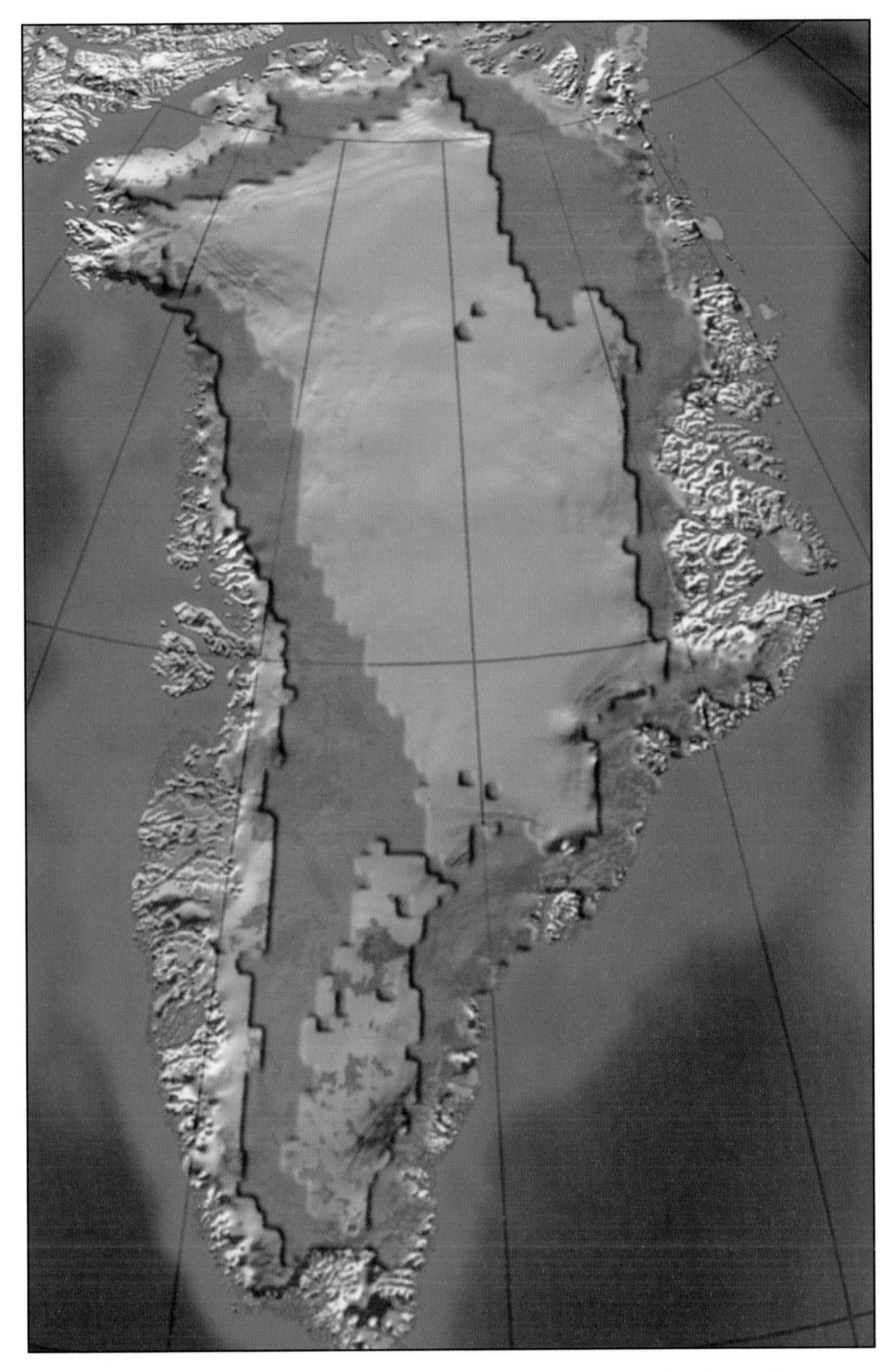

The second image released by the ACIA shows the Greenland Ice Sheet in 2002. The increase in orange represents how much ice has decreased in the 10 year period.

The rise in sea levels from ice melt and from natural expansion of water as it warms places cities and communities around the world at risk of increased yearly flooding. Researchers at the University of Bristol in Britain published a study identifying certain areas of the world, including parts of tropical Africa and northwest South America, that are particularly susceptible to global warming–induced floods. The British government has been concerned about the risk of flooding, especially along the Thames River. In 2004 the government's chief scientific adviser, David A. King, wrote in the American magazine *Science* "in the worse-case scenario, the number of people at 'high' risk of flooding in Britain will more than double to nearly 3.5 million. Potential economic damage to properties runs into tens of billions of pounds per annum."[1] It is important to note, however, that King is identifying a potential scenario, not one that has actually occurred.

The rise in sea levels from ice melt and from natural expansion of water as it warms places cities and communities around the world at risk of increased yearly flooding.

Increased Risk of Drought and Wildfire

On the opposite end of the spectrum, increased global temperatures also bring increased risk of drought and wildfires. John P. Holdren, president of the American Association for the Advancement of Science (AAAS) and Alan I. Leshner, publisher of *Science*, wrote in 2006 that "[increased] fire threat is linked to warming temperatures and earlier snowmelts . . . [and] wildfires have been increasing all across the globe."[2] Fires are a direct result of increased drought. By running a simulated climate model, David Rind of the Goddard Institute for Space Studies has found that as carbon dioxide levels rise, Earth could begin experiencing increased drought conditions and water shortages starting near the equator and heading toward the poles, caused by the drier soil that increases with higher temperatures. If Rind runs his model using carbon dioxide levels that are double the preindustrial levels, most of the United States is shown under drought.

Drought caused by climate change has been a particular concern for Australia. The country has a naturally arid climate and is one of the world's largest polluters after the United States. Since 2001 Australia has experienced below-average rainfall, leading to severe drought conditions. Australian science writer Mark Lynas states that he believes that the country may have passed the point of no return. In the *New Statesman*, he writes,

> I now think a tipping point has also been crossed in Australia, pitching the continent into a permanently drier climate regime that will wipe out most of its agricultural base and leave its cities constantly threatened by water shortages, heatwaves and uncontrollable wildfires . . . this is the scale of drought we're talking about here. Australia's new climate will be different from anything ever experienced since humans first settled on the continent.[3]

Economic and Social Effects

All of these events have an effect not only on the natural systems of Earth but also on human civilization. Public resources are strained during times of natural disasters. Resources would be stretched much thinner if these events were to become more frequent and more severe. In addition, people could experience loss of property, jobs, and businesses as areas are submerged by flooding or destroyed by hurricanes or fires. Communities could potentially break up as people move from area to area to avoid these events as well as the accompanying disease and health problems.

Scientists have theorized that Earth is reaching the end of the Holocene geological epoch, which began 8,000 years ago. This peaceful era has been so pleasant and stable that it has sometimes been called "The Long Summer." A new era may be beginning. Some have proposed that it be named the Anthropocene epoch—the only geological age to be defined by human beings. Some commentators think that human beings have become capable of permanently altering Earth's climate and atmosphere.

What Are the Controversies Surrounding Global Warming?

While the issue of global warming may seem to be a straightforward scientific matter, like the existence of the moon or photosynthesis, it is actually nothing of the sort. Global warming and climate change are matters of vigorous public debate that are far from resolved. Some scientists, policy makers, and members of the public have not committed fully to the idea that global warming is real, caused by humans, and an urgent situation. In contrast, many in the scientific community have declared that all three of the above statements are true. In 2003 the American Meteorological Society issued this statement as part of a report on climate change: "The theory of how greenhouse gases directly interact with atmospheric radiation is not controversial. If no other factors counter their influence, increases in their concentration will lead to global warming."[4]

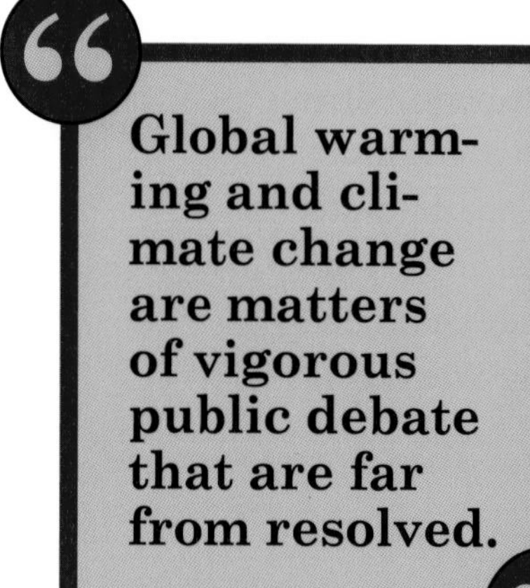

Other scientists, while supporting the theory that global warming is real and caused by human activity, also warn against sensationalizing its potential effects. In March 2007, Paul Hardaker, a professor and scientist with the British Royal Meteorological Society, told the BBC that he fears the issue will not be taken seriously if it is overblown. He stated,

> Organizations have been guilty of overplaying the message. . . . I think we do have to be careful as scientists not to overstate the case because it does damage the credibility of the many other things we have greater certainty about. . . . We have to stick to what the science is telling us; and I don't think making that sound more sensational . . . is the right thing for us to be doing.[5]

What Are the Solutions for Global Warming?

For the most part, the world's governmental bodies have recognized the need to mitigate carbon dioxide emissions and reduce dependence on carbon-producing systems. The United Nations has taken climate change seriously. In 1988 the UN established the Intergovernmental Panel on Climate Change (IPCC) for the purpose of studying what it calls "human-induced climate change." The panel assesses all of the major scientific research on climate change and supports the activities of the United Nations Framework Convention on Climate Change (UNFCCC). The IPCC is regarded as the authoritative source on climate change issues. The UNFCCC and the IPCC were the authors of the Kyoto Protocol.

To date, the Kyoto Protocol is the most comprehensive initiative seeking to cut greenhouse gas emissions on a global scale. An amendment to the UNFCCC, the Kyoto Protocol requires that all countries that have ratified the agreement cut their carbon emissions to certain levels, depending on the country, by certain dates. The treaty has been signed by 169 of the world's countries, including the United Kingdom, the European Union, Mexico, Canada, virtually all of the countries in South America, and most of the former Soviet Union. The United States has not signed the Protocol.

When Russia ratified the treaty in 2004, the Kyoto Protocol achieved coverage of 55 percent of the levels it set and was able to enter into force under the agreement. At the time of the Russian ratification, the executive secretary of the UNFCCC stated, "The entry into force of the Kyoto Protocol will be an important milestone in the history of international efforts to protect the earth's climate from human interference."[6]

Global Efforts

Of the world's industrialized countries, the United Kingdom has taken the most proactive stance on global warming. Aside from ratifying the Kyoto Protocol, the country has taken further voluntary steps to curb its greenhouse gas emissions. In 2006 it commissioned *The Stern Review on the Economics of Climate Change*, a major study of the effects global warming will have on the world community. Moreover, prime minister Tony Blair and his likely successor, Gordon Brown, have pledged to cut Britain's carbon emissions by 60 percent over the next 40 years.

The U.S. refusal to ratify the Kyoto Protocol, most people argue,

renders the agreement virtually useless; the nation is the world's single-largest producer of greenhouse gases. In addition, the federal government has no official system in place to curb greenhouse gas emissions. If and when the federal government does decide to take wide-scale, long-range action on global warming, it is unlikely to introduce a tax on carbon emissions. A more likely plan is the free-market cap-and-trade system, in which companies have a limit on how much carbon they are permitted to produce in a year. If they are under their maximum at the end of the year, they can sell credits to other companies. In essence, this is the system put in place by the Kyoto Protocol, in which countries have a mandatory carbon limit and can buy and sell carbon credits from other countries.

State and Local Efforts

Frustrated by the U.S. federal government's unwillingness to ratify the Kyoto treaty or take a stance on global warming, some states have begun to pass their own caps on greenhouse gas emissions. Seven northeastern states have formed the Regional Greenhouse Gas Initiative, for instance, imposing caps on emissions from electric power. Oregon has passed laws mandating that emissions levels be cut to 75 percent below 1990 levels by 2050. Altogether 22 states have voluntarily established programs requiring that a percentage of their electricity come from renewable, noncarbon energy sources.

The efforts are not limited to states. The U.S. Mayors Climate Protection Agreement, an initiative to reduce greenhouse gas levels to the levels set by the Kyoto Protocol, has been signed by 279 cities around the country. Those involved include Seattle; New York City; Washington, D.C.; Chicago; Los Angeles; San Francisco; and Philadelphia.

Alternative Energy

Alternative energy sources are the keystone to curbing or eliminating greenhouse gas emissions. Many of these technologies, however, are either in the beginning stages of development or are not widely available to the general population. Some, especially newer sources, are expensive compared to coal, oil, and gas power.

Alternative sources of energy include solar power, wind power, landfill gas, geothermal power, biomass power, hydrogen and fuel cells, and hydropower, though this list is far from inclusive. All of these sources share the distinction of being renewable—meaning that they do not rely on a finite source—and most are either noncarbon emitting or produce only a small amount of carbon. Yet each alternative energy source is controversial because of efficiency and cost issues. Wind power, for instance, is very clean; the turbines emit zero carbon. In addition, one wind farm can produce enough electricity to power 24,000 homes. This is the equivalent of burning 50,000 tons (45,359t) of coal to get the same amount of electricity. Wind technology is advanced compared to other alternative sources, making it affordable. On the other hand, wind farms take up large tracts of land, and some people find them unattractive. Only some locations produce enough wind. Constructing the actual wind farm involves clearing land of trees, leveling the earth, and sinking huge concrete and metal pillars—all ecologically invasive.

> **“Yet each alternative energy source is controversial because of efficiency and cost issues.”**

No one alternative energy source is ideal, but economist Nicholas Stern, author of the *Stern Review*, points out that people will have to make up their minds: In order to cut greenhouse gas emissions significantly, 60 percent of global power will have to come from noncarbon sources by 2050. Presently, these sources supply only 14 percent of the world's power.

The climate change debate touches a nerve in the global consciousness. At issue are some of the most fundamental human needs: clean water, safe shelter, clean air, and a secure future for children. The systems—power, transportation, agriculture, business—on which modern civilization was built are being questioned. *Compact Research: Global Warming and Climate Change* will attempt to shed light on this important and contentious debate.

What Is Global Warming and Climate Change?

“Global warming is one of the most serious challenges facing us today.”

—Union of Concerned Scientists.

“[Let’s] consider the possibility that a significant portion of the warming may be natural, and that regions . . . are likely to have experienced as warm or warmer conditions in their climate history.”

—*World Climate Report*, January 8, 2007.

Breakdown of Various Greenhouse Gases

A greenhouse gas is defined as any gas that permits light from the sun to enter Earth’s atmosphere but traps some of the sun’s radiant heat, rather than allowing that heat to reflect back into space. The trapped heat then warms the atmosphere. There are about 30 greenhouse gases all together. Carbon dioxide tends to get the most attention; 80 percent of all greenhouse gas is carbon dioxide. Other gases are not as well known: methane, nitrous oxide, sulfur hexachloride, perfluorocarbons (PFCs), and hydrofluorocarbons (HFCs). Methane and nitrous oxide occur naturally in the atmosphere, while sulfur hexachloride, PFCs, and HFCs are entirely man-made. Sulfur hexachloride and PFCs are by-products of aluminum smelting and electricity production. HFCs are used in refrigeration systems.

In small amounts these gases are mostly harmless. But even those that occur naturally are now being produced in a ratio that is higher than any in existence previously. Methane gas, for example, exists naturally in Earth's crust, in animals, and in the breakdown and decay of organic matter. It is a necessary, useful gas. However, large amounts of methane are being produced by humans—in landfills, wastewater treatment, and livestock farming—out of the natural proportion.

Interestingly, water vapor is also a greenhouse gas. It is not a pollutant, which differentiates it from its fellows, but it does contribute to climate change—though not necessarily to global warming. Science writer Tim Flannery explains in his book *The Weather Makers* that clouds (made up of water vapor) can both reflect light and trap heat. High, thin clouds tend to have a warming effect on the atmosphere, while low, thick clouds have a cooling effect. Water vapor also exists in a self-perpetuating cycle: Evaporation increases as heat increases, leading to more precipitation, which in turn leads to more evaporation.

Natural Causes of Global Warming and Climate Change

When most people hear about "climate change" they immediately think of cars massing on highways and power plants spewing smoke, all pouring carbon dioxide into the atmosphere. But some people know that climate change does not always mean "hotter"—sometimes, it can mean "cooler." Moreover, climate change can be caused by natural events on Earth. In a policy paper, the American Meteorological Society writes, "Prior to the industrial age, natural processes . . . were the dominant . . . factors producing long-term climate changes over periods of decades, centuries, and millennia."[7]

> **"But even those [gases] that occur naturally are now being produced in a ratio that is higher than any in existence previously."**

Volcanic eruptions, for instance, have historically been a factor in natural global warming. In her book *Field Notes from a Catastrophe* Elizabeth Kolbert explains that eruptions release enormous amounts of sulfur dioxide, a greenhouse gas, into the atmosphere in the form of tiny aerosol droplets. The droplets remain in the atmosphere, reflecting sunlight back into space, away from Earth, thus cooling it.

The sun's intensity also changes naturally over time—this is called "sunspot activity." As the heat and light from the sun increases or decreases, even slightly, it can greatly affect temperatures on Earth. Some climate scientists have speculated that sunspot activity may have brought on the last cooling period in the Western Hemisphere, known as the Little Ice Age, which spread ice and snow across Europe from the 16th to the 18th centuries.

> **"Climate change can be caused by natural events on Earth."**

Natural climate change does not occur only in the atmosphere, however. The oceans—a major factor in determining Earth's temperatures as well as sustaining human life—are also subject to a natural cycle of warming and cooling. Scientists have used the preserved bodies of tiny ocean organisms to determine that about every 1,500 years the ocean's temperature becomes much cooler. This trend persists for a few centuries, and then temperatures rise again. One of these cooler cycles coincided with the Little Ice Age.

Carbon Dioxide

Understanding the role of carbon dioxide is crucial to the understanding of climate change. Of all greenhouse gases in the atmosphere today, 80 percent is carbon dioxide. Burning oil, natural gas, and coal; burning forests; and cement production all release carbon dioxide into the atmosphere. Though detractors exist, most scientists now agree on two key points: that carbon dioxide in large amounts helps to warm Earth and that the excess of carbon dioxide in the atmosphere today is a result of human activity—namely, industrialization and the invention of coal-burning devices.

Carbon dioxide is measured using the common scientific notation of "parts per million." Ppm expresses the concentration of carbon in the air. A level of one ppm would mean one unit of carbon for every million units of air. Carbon can also be measured in pounds or in tons. To understand the rate at which carbon dioxide levels have changed over time, scientists compare preindustrial and current levels.

Measuring climate change, including global warming, is only meaningful if it can be compared with historical information. Modern climate change measurements have existed only for the last half-century, yet scientists have found a way to unlock the secrets of the weather that are imprisoned in Earth itself.

Unlocking Past Climate Change

The massive glaciers all over the world offer a record of what climate change was like in the past. Over the millennia, snow that has fallen on these glaciers has never melted. It rests on the ice and each year is covered with more snow. The ice crystals remain intact, layer upon layer. Nearer the top, the layers are thick and fluffy. As they continue down, they are compressed, getting thinner and denser. Contained within the ice is the prize: tiny air bubbles from the year the snow fell, even as far back as 100,000 years.

Scientists have developed technology that allows them to drill deep into these ice sheets and extract miles-long cylinders of ice, called ice cores. Greenland has particularly valuable ice. Since 1990 three Greenlandic ice cores have been extracted, each nearly 2 miles (3.2km) long. By examining the air in the bubbles in these cores, much as one could examine the rings on a tree, scientists can obtain a record of just how much carbon dioxide was in the air at any given time, as well as calculate the temperature of the air that year. This has enabled scientists to understand how the climate has changed over the millennia.

> **"Of all greenhouse gases in the atmosphere today, 80 percent is carbon dioxide."**

In 1958 scientists began some of the first official modern measurements of global warming. They set up equipment on the slopes of the Mauna Loa volcano in Hawaii and measured carbon dioxide levels there every year. The graph they created is well known in climate circles. It is called the Keeling Curve, after Charles Keeling, the scientist who first measured the steep upward path. Using the Mauna Loa measurements, among others, scientists have estimated that preindustrial carbon levels were about 225 parts per million. Kolbert writes that as of the summer of 2005, carbon levels were at 378 ppm and expected to reach 500 ppm, double preindustrial levels, by the middle of the 21st century.

Burning coal is a cheap, effective, and—as of the start of the twenty-first century—abundant way to generate electricity. It is also dirty: Depending on the type, the composition of coal ranges from 70 to 90 percent carbon. Yet the technology of many coal-fired power plants has changed very little since it was first invented in the nineteenth century: The coal is burned to boil water, which creates steam, moving turbines that then generate electricity. Some power plants can burn 550 tons (499t) of coal per hour.

A few plants, especially in the United States and Europe, have pollution controls, including filters and scrubbers, meant to remove some of the carbon dioxide before it is released into the atmosphere. But in developing countries, particularly China and India, where energy needs are rapidly increasing, coal-fired plants are being built at a rapid pace, using mostly antiquated parts and technology—without modern and expensive pollution controls.

Man-Made Causes: Deforestation and Transport

For most Americans transportation is the cause of global warming that is most familiar, aside from electricity. Most American families own at least one car. Every day, hundreds of thousands of tractor-trailers, giant container ships, and freight trains cross the country and the globe—all burning petroleum gasoline. Gas, especially diesel gas, produces carbon dioxide as well as carbon monoxide and other pollutants. In fact, 1 gallon (3.8L) of gasoline produces 5 pounds (2.3kg) of carbon. In the United States, transportation needs contribute 32 percent of carbon emissions.

> **"Scientists have found a way to unlock the secrets of the weather that are imprisoned in Earth itself."**

Cities in developing nations, such as Beijing, New Delhi, Calcutta, and Seoul, have far higher levels of carbon dioxide emissions from cars and trucks than the United States, due to widespread use of diesel engines and outdated technology. The *International Herald Tribune* reports that in Asian cities, 30 to 70 percent of all pollution comes from motor vehicles. The *Tribune* cites a list compiled by the Asian Development Bank which found that Beijing had air pollution levels that were more than six times that of Paris, London, or New York.

Airplanes as a source of carbon emissions have been considered only recently. According to an article in *USA Today*, British climate scientists recently predicted that by 2050, aircraft emissions will be one of the biggest contributors to global warming, when the rapid rise in global air travel is factored in. Jet fuel exudes carbon dioxide, among other greenhouse gases, while on the ground, but scientists are particularly concerned about the emissions when the planes are flying. High in the atmosphere, the effects of carbon dioxide are amplified and also tend to remain longer in the atmosphere, already a vulnerable area.

> **"Some power plants can burn 550 tons (499t) of coal per hour."**

Deforestation is the cause of global warming that has been the most overlooked, though it contributes 18 percent of the world's carbon emissions. As most elementary school children know, trees are natural absorbers of carbon dioxide—they breathe in the gas and exhale oxygen. As the world's forests are cut down either for their resources, to make way for development, or to clear land for agriculture, they are replaced by structures that do not perform the same function. In addition, the methods used to clear forests often add their own emissions: Slash-and-burn clearing or clear-cutting of trees pours smoke into the atmosphere, as do the trucks and bulldozers needed to clear the harvested trees. Each year 30 million acres (12 million ha) of forest are cleared, according to the Food and Agriculture Organization, a division of the UN. The affected forests are almost entirely tropical, the FAO reports, with the problem worst in Africa, Latin America, and Southeast Asia.

Albedo

Albedo is a term rarely heard outside discussions about climate change and global warming, yet it is also crucial to understanding the issue. Albedo is the amount of light and heat reflected into the atmosphere from Earth's surface. White areas reflect more light than dark areas. Ice, such as icebergs and glaciers, is a white surface, as are snow cover and grasslands. The more of these surfaces that exist on Earth, the more light and heat is reflected away from Earth's surface rather than remaining and heating it up. Dark surfaces include water, soil, and forests. These tend to absorb and retain light and heat. Earth exists in a balancing act between its light-reflecting and heat-absorbing surfaces. With the present state of climate change, however, that balance has been drastically altered.

Within the light/dark classifications, water absorbs huge amounts of heat, ice practically none. As sea and land ice melt from increased temperatures, they are replaced by heat-retaining water. Areas of Earth that have been permanently covered with ice and snow, such as Siberia and parts of northern Canada, are now exposing their dark topsoil as their snow cover melts. Soot from forest clearing, wildfires, and the burning of fossil fuels also contribute to the imbalance as it is carried aloft, covering existing ice and snow and reducing their albedo.

"High in the atmosphere, the effects of carbon dioxide are amplified and also tend to remain longer in the atmosphere, already a vulnerable area."

Reforestation efforts are generally seen as positive, but ironically, some reforestation can negatively impact albedo. When forests, which are dark and thus heat-absorbing, are planted on what was formerly grassland, the albedo of the area is altered. Climate scientists recommend only reforesting areas that were already forestland before.

To further complicate matters, the effect is reversed in tropical forests. In that particular climate, the presence of the plants encourages the formation of low, heavy cloud cover, a white, heat-reflecting surface.

The Pace of Climate Change

Kolbert points out that during the first decade of the 21st century, Earth is as warm as it has been in the last 42,000 years. The last time the Earth was this hot, Kolbert writes, was 3.5 million years ago during an age called the Pliocene. Carbon dioxide levels have not been higher since the Eocene, 50 million years ago, when there were crocodiles in Colorado. Modern humans have never lived in an atmosphere as extreme as ours is now. The ice cores have revealed that it will be even harder than previously thought to predict what climate will be like in the future.

For most of modern human history, people have believed that Earth's climate changed slowly over the course of centuries or even millennia. However, armed with the information from the ice cores, scientists now agree that Earth is capable of sudden radical shifts in temperature, precipitation, and wind patterns, among other factors. Some believe that climate change was responsible for the demise of entire civilizations, such as the Norse Vikings or the Mayans of South America.

One particular proponent of this theory is environmental journalist Eugene Linden. Linden explains that weather over the long term can be unpredictable, changing rapidly over the course of only a few decades. He writes that 11th century Vikings left Iceland and Greenland to explore the New World during a particularly warm period called the Medieval Warming. They could have prospered in North America except that the weather began to change, turning colder during the 12th century in a period now called the Little Ice Age. The actual temperature difference, Linden estimates, was only about two degrees, but this was enough to increase sea ice, spur glaciers to move, and impact the food supply. The Viking settlements could not survive. The Little Ice Age eventually forced the Vikings to abandon their settlements in Greenland also, though they had prospered there for hundreds of years.

> **Earth exists in a balancing act between its light-reflecting and heat-absorbing surfaces.**

Abrupt climate change may also have contributed to the demise of the Mayan civilization that existed in Central America from approximately 1000 B.C. until about A.D. 900. Researchers have found evidence that the Yucatan Peninsula experienced a period of severe drought from about A.D. 750 to 900, brought on by natural variations in the intensity of the sun. This drought, the worst in 7,000 years, coincided with the decline of the Mayan civilization.

Linden points out that all of the warmest years on record have been since the 1980s and during the 1990s. The pace at which temperature records are being broken has accelerated. The high carbon dioxide levels combined with our new knowledge about how quickly climate can change only make the future more unpredictable.

Primary Source Quotes*

What Is Global Warming and Climate Change?

"The Earth's climate system has demonstrably changed on both global and regional scales since the pre-industrial era, and there is new and stronger evidence that most of the warming observed . . . is attributable to human activities."

—Rajendra Pachauri, "Address by Dr. R.K. Pachauri, Chairman of the Intergovernmental Panel on Climate Change (IPCC) at the High Level Segment, Montreal, Canada," 11th Conference of the Parties to the United Nations Framework Convention, December 7, 2005.

Pachauri is the chairman of the Intergovernmental Panel on Climate Change.

"Climate changes naturally all the time, partly in predictable cycles, and partly in unpredictable shorter rhythms and rapid episodic shifts, some of which the causes remain unknown."

—Bob Carter, "There IS a Problem with Global Warming. . . . It Stopped in 1998," *London Telegraph*, September 4, 2006.

Carter is a geologist at James Cook University in Queensland, Australia. He specializes in paleoclimate research.

Bracketed quotes indicate conflicting positions.

* Editor's Note: While the definition of a primary source can be narrowly or broadly defined, for the purposes of Compact Research, a primary source consists of: 1) results of original research presented by an organization or researcher; 2) eyewitness accounts of events, personal experience, or work experience; 3) first-person editorials offering pundits' opinions; 4) government officials presenting political plans and/or policies; 5) representatives of organizations presenting testimony or policy.

"Global warming can occur from a variety of causes, both natural and human induced. In common usage, 'global warming' often refers to the warming that can occur as a result of increased emissions of greenhouse gases from human activities."

—Environmental Protection Agency, "Climate Change or Global Warming?" http://epa.gov.

The Environmental Protection Agency is an agency of the federal government in charge of protecting the air, water, and land of the environment.

"The oldest geologic sediments suggest that, before life evolved, the concentration of atmospheric carbon dioxide may have been one hundred times that of the present, providing a substantial greenhouse effect."

—Earth Observatory, "The Carbon Cycle," http://earthobservatory.nasa.gov.

The Earth Observatory is a division of NASA and provides public information about Earth taken from satellite images.

"Anthropogenic carbon dioxide comes from fossil fuel combustion, changes in land use (e.g., forest clearing), and cement manufacture."

—Carbon Dioxide Information Analysis Center, "Frequently Asked Global Change Questions," http://cdiac.ornl.gov.

The Carbon Dioxide Information Analysis Center is a data and information center for the U.S. Department of Energy.

"There is no previous human experience of Earth's atmosphere at current levels of greenhouse gases to assist us to predict the outcomes. It is likely, though, that the natural oscillating pattern of ice ages and warm periods is now being disturbed."

—David King, "Dealing with Climate Change: The Scope of the Challenge," Climate Change: Meeting the Challenge Together, Berlin, November 3, 2004.

King is the United Kingdom's chief scientific adviser.

"What serious dispute there is about climate change relates mostly to the distinction between natural change and human-driven change. . . . [The] human impact is now becoming unmistakably evident."

—Crispin Tickell, "Vulnerable Earth," Robert C. Barnard Environmental Lecture, American Association for the Advancement of Science, Washington, D.C., September 18, 2006.

Tickell is a diplomat and environmental climate change scholar. He is the director of the Policy Foresight Program at the University of Oxford, England.

"[We] are heading over the next several generations into a climate that has never before been experienced by civilization. Changes in temperature, precipitation, sea level, and weather extremes will affect many aspects of human life."

—Rick Anthes, "Strange Bedfellows and Holy Alliances," *UCAR Quarterly*, Fall 2006.

Anthes is an atmospheric scientist and the president of the University Corporation for Atmospheric Research.

"The observed patterns of change over the past 50 years cannot be explained by natural processes alone, nor by the effects of short-lived atmospheric constituents (such as aerosols and tropospheric ozone) alone."

—Tom L. Wigley, "Trends in the Lower Atmosphere: Steps for Understanding and Reconciling Differences," *U.S. Climate Change Science Program: Synthesis and Assessment Product*, April 2006.

Wigley is a meteorologist and senior scientist at the National Center for Atmospheric Research in Boulder, Colorado.

"Primary energy supply is on the threshold of a revolution as profound as utilization of carbon-based fuels by the Industrial Revolution."

—Alistair Miller and Romney B. Duffey, "Integrating Large-Scale Co-generation of Hydrogen and Electricity from Wind and Nuclear Sources," Climate Change and Technology Conference, Ottawa, Canada, May 10–12, 2006.

Miller is a chemical engineer and the former president of the Canadian Society for Chemical Engineering. Duffey is the principal scientist of Atomic Energy of Canada.

"The year 2005 was most likely one of the two warmest years on record since 1850. Arctic sea ice during September 2005 was the lowest on record."

—Michel Jarraud, foreword to *WMO Statement on the Status of the Global Climate in 2005*, 2006.

Jarraud is the secretary-general of the World Meteorological Organization, part of the United Nations.

“Most of the driving forces causing the degradation of ecosystems are either staying constant or growing in intensity, and two—climate change and excessive nutrient loading —will become major drivers of change in the next 50 years.”

—Walter Reid, “Presentation by Dr. Walter Reid at the UN Permanent Forum on Indigenous Issues,” United Nations Permanent Forum on Indigenous Issues, New York, May 18, 2005.

Reid is a zoologist and the director of Millennium Ecosystem Assessment, a UN subgroup that studies human-induced ecosystem change.

“A well-documented rise in global temperatures has coincided with a significant increase in the concentration of carbon dioxide in the atmosphere. Respected scientists believe the two trends are related.”

—John Paul Stevens, *Massachusetts et al., Petitioners, v. Environmental Protection Agency et al.*, opinion of the Court, Supreme Court of the United States, April 2, 2007.

Stevens is a U.S. Supreme Court justice.

What Is Global Warming and Climate Change?

- According to the National Oceanic and Atmospheric Administration, without a natural greenhouse effect the temperature of Earth would be about **0°F (–18°C)** instead of its present 57°F (14°C).

- NASA's Goddard Institute for Space Studies states that surface temperatures on Earth have increased by **0.36°F (0.2°C)** every decade for the last 30 years.

- According to the American Geophysical Union, **80 percent** of the increase in carbon dioxide in the atmosphere since the 1700s has occurred during the 20th and 21st centuries.

- Emissions in 2006 rose **twice** as fast as in 2000, according to the Global Carbon Project.

- The Union of Concerned Scientists states that every year since 1992 is on the list of the **20 warmest years** on record.

- According to the Pew Center on Global Climate Change, the 1990s were the hottest decade of the last **150 years**.

- Reduced solar activity from the 1400s to the 1700s was likely a key factor in the **"Little Ice Age"** that resulted in a slight cooling of North America, Europe, and probably other areas around the globe, according to the EPA.

Understanding the Greenhouse Effect

The following illustration depicts the greenhouse effect, the process by which light and heat from the sun are either absorbed by the earth's surface or are reflected back into space. The composition of the earth's atmosphere is key in determining how much light and heat are absorbed and reflected.

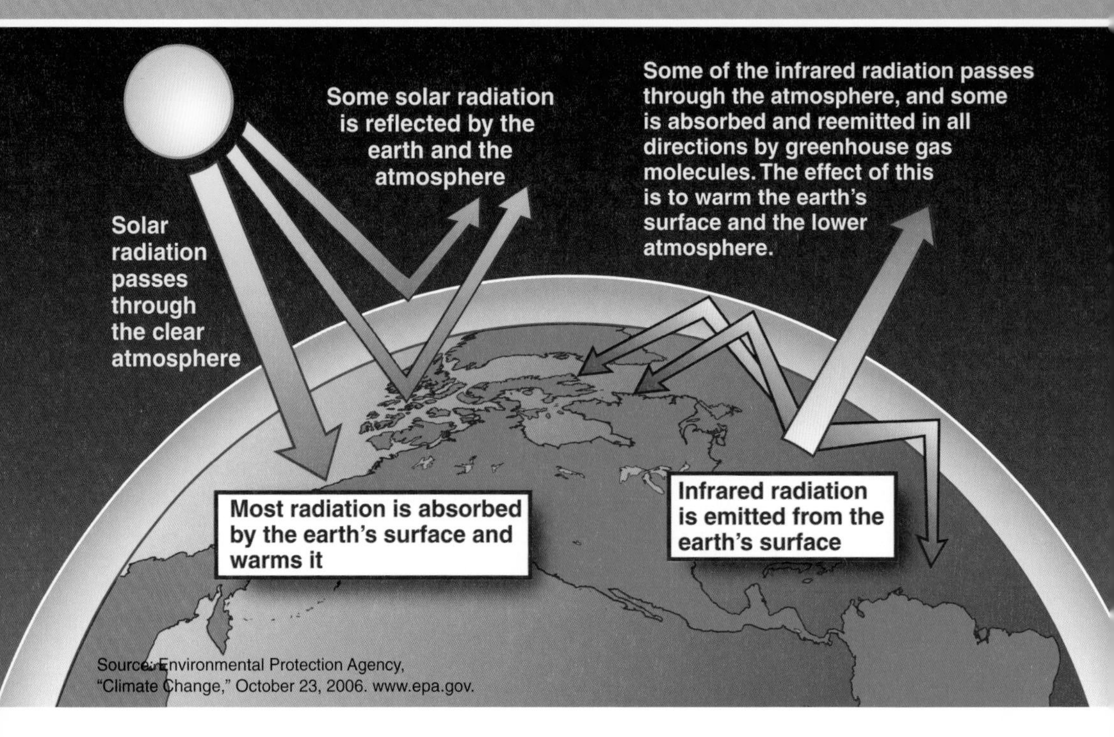

Source: Environmental Protection Agency, "Climate Change," October 23, 2006. www.epa.gov.

- Jet engines account for **3 percent** of all carbon dioxide that contributes to global warming, according to *USA Today*.
- According to the Soil Association of Britain, conventional agriculture contributes about **8 percent** of greenhouse gas emissions.
- According to NASA's Earth Observatory, ice core samples taken in Antarctica and Greenland have shown that carbon dioxide concentrations during the last ice age were only **50 percent** of what they are today.

- About **30 percent** of the total solar energy that strikes Earth is reflected back into space by clouds, atmospheric aerosols, reflective ground surfaces, and ocean surf. The remaining **70 percent** is absorbed by the land, air, and the oceans, states NASA.

- According to the Intergovernmental Panel on Climate Change, as carbon dioxide continues to rise, temperatures will rise between **three and five degrees** by the end of the century.

Carbon Dioxide Levels from the Mauna Loa Observatory

Carbon dioxide levels have been recorded at the historic Mauna Loa Observatory in Hawaii since 1960. According to this chart from the National Oceanic and Atmospheric Administration, levels are still on the rise, with 383 parts per million in January 2007.

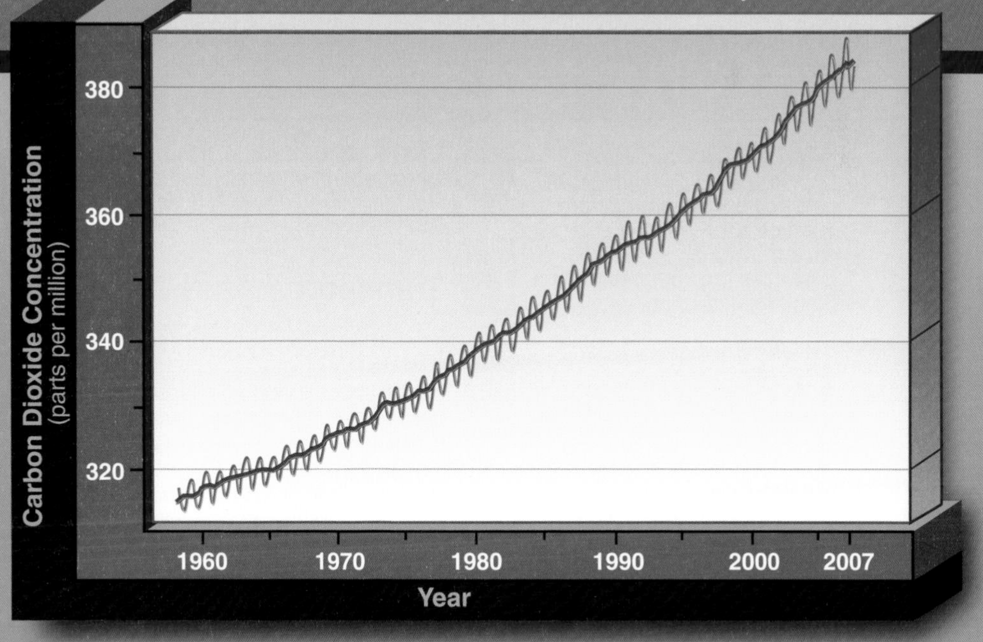

Source: National Oceanic and Atmospheric Administration, Earth System Research Laboratory, January 2007. www.cmdl.noaa.gov.

Breakdown of Global Greenhouse Gas Emissions by Sector

According to this graph from the *Stern Review*, a British study of global warming, power production is the largest contributor to global emissions, at 24 percent. The second largest contributor is land use, at 18 percent.

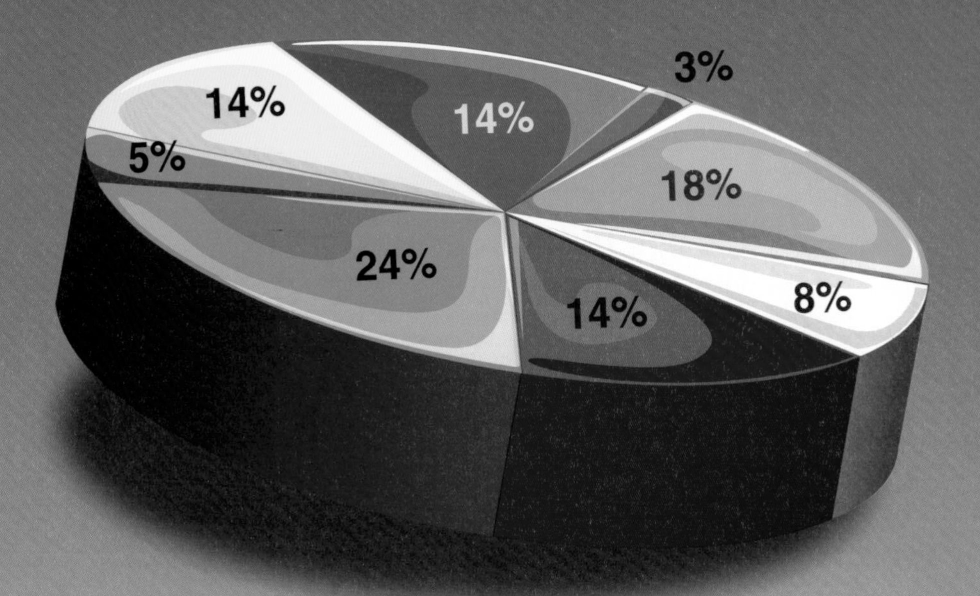

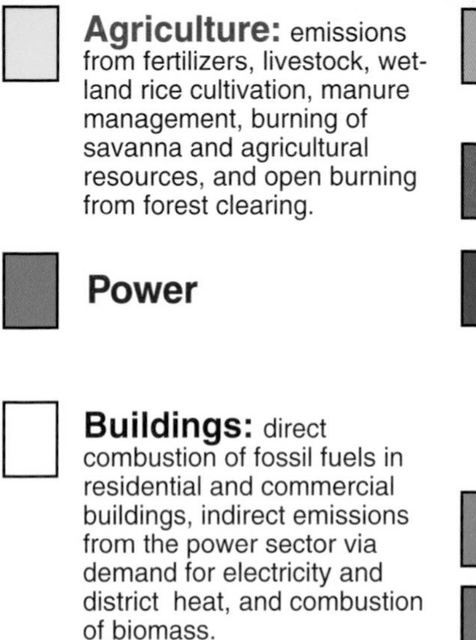

Land use: change in the management of land by humans, such as conversion of forests to pasture land.

Transport

Industry: direct CO_2 emissions from manufacturing and construction, from chemical processes involved in producing various chemicals and metals, "upstream" emissions from the power sector, and from transport, indirectly from the movement of goods.

Other

Waste

Source: Nicholas Stern, "Global Emissions by Sector," *Stern Review on the Economics of Climate Change*, October 30, 2006. www.hm-treasury.gov.uk.

Rise in Carbon Dioxide and Emissions from Energy Production

Over the next few decades, if current energy production remains the same and no new policies are implemented, worldwide carbon dioxide emissions from fossil fuels will rise to 40 billion tons from the 1990 level of just over 20 billion tons. Claude Mandil of the International Energy Agency notes that half of the increase will come from new coal-fired power plants, primarily in China and India.

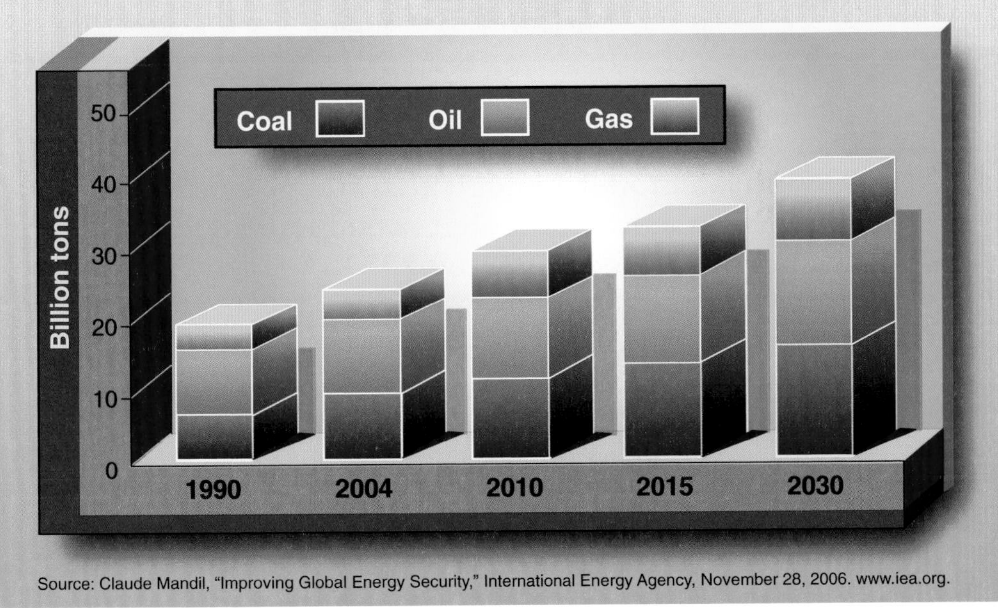

Source: Claude Mandil, "Improving Global Energy Security," International Energy Agency, November 28, 2006. www.iea.org.

- As of April 20, 2007, **10 states** had banded together to voluntarily form the Northeast Regional Greenhouse Gas Initiative.
- According to the Pew Center on Global Climate Change, **coastal ecosystems** are particularly sensitive to the effects of global warming.
- Humans first began producing large amounts of greenhouse gases during the **Industrial Revolution** of the eighteenth and nineteenth centuries.

What Are the Consequences of Global Warming?

"Observed changes in regional climate have affected many physical and biological systems, and there are preliminary indications that social and economic systems have also been affected."

—Rajenda Pachauri, 11th Conference of the Parties to the United Nations Framework Convention on Climate Change.

"[A] warmer planet has beneficial effects on food production. . . . Global warming also increases carbon dioxide, which acts like fertilizer for plants."

—Dennis T. Avery, "Global Warming—Famine or Feast?"

Melting of Sea Ice

The state of the word's sea ice is of concern to climate scientists. As Earth warms, the massive ice sheets that float in the Arctic and Antarctic seas are melting at a rate faster than scientists previously anticipated. The *U.S. Naval Institute Proceedings* noted in January 2007 that since 2002, "[winter] ice cover is now decreasing at the rate of more than 3 percent a decade. . . . Since the 1970s, summer ice cover has decreased by 30 percent while the total volume of ice decreased by about 40 percent."[8] If the present rate of melt continues, the Arctic Ocean could potentially be free of ice during the summer in 10 to 50 years.

The *Proceedings* notes that this change could be beneficial, stating, "[Greatly] reduced ice cover will permit global maritime shipping to use [the Arctic] ocean for safe navigation in all but the coldest winter months. Compared to ship routes used today, a northern sea route between Asia and Europe would save 4,000 miles in the transport of goods between the two regions."[9] The *New York Times* reports that an entrepreneur, anticipating the possibilities of the opened shipping routes, bought the Port of Churchill in Hudson Bay in 1997 for the sum of seven dollars. The port is anticipated to be a key point in the new route from Russia to North America.

The consequences of sea ice melt are difficult to predict, though, as underscored by the collapse of the Larsen B ice shelf in 2002. This massive ice mass off the coast of the Antarctic Peninsula was 700 feet (213m) thick, 150 miles (241km) long and 30 miles (48km) wide—the size of Luxembourg. In February 2002 it cracked for the first time in recorded history. Over the course of the next 35 days, it broke up and disappeared completely.

Sea ice melt, like Larsen B, does not raise sea levels, since the ice was already floating in the ocean. Scientists found, however, that glacier melt, which does increase sea levels, was now flowing more rapidly. Larsen B, sea-based ice, had acted like a dam to the land ice. When the sea ice broke up, the land ice behind it shifted, increasing melt and causing chunks of glacier to fall into the ocean. Meltwater and broken ice does raise global sea levels, as well as affect the composition of ocean water.

> **"If the present rate of melt continues, the Artic Ocean could potentially be free of ice during the summer in 10 to 50 years."**

However, it is important to note that the breakup of Larsen B and other sea ice sheets do not signal catastrophe. Rather, as John Young, an environmental consultant writes in *World Watch* magazine, "[Melting of polar ice] makes catastrophic developments conceivable. Until now, global climate models have generally assumed that melting of ice in polar regions would take thousands of years. New scientific developments suggest that sea levels will rise more quickly . . . than the IPCC predicted [in 2001]."[10] Another concern is that at some unknown point, a threshold will be crossed beyond which the melting of sea ice sheets will become impossible to stop, even if ocean warming is halted. These theories, however, remain predictions and concerns for the future, not present realities.

The Importance of Greenlandic Ice

Greenland, a northern landmass covered almost entirely by glaciers, has been of particular concern to climate scientists. The region holds an enormous amount of snow and ice: 10 percent of the world's land ice mass, according to *New Scientist*, is contained within the world's second-largest icecap. The area of the entire Greenland ice sheet is 1.09 million square miles (2.83 million sq. km), the same size as mainland Europe. Increased global temperatures have caused the Greenlandic ice, which is land-based, to melt. Since the 1990s the Greenland ice sheet has been shrinking at the rate of 12 cubic miles (50 cu. km) per year. As the glacier ice melts, meltwater gathers at the base and lubricates the ice, which speeds up the melt rate.

> **"However, it is important to note that the breakup of Larsen B and other sea ice sheets do not signal catastrophe."**

In March 2006 scientists at the National Center for Atmospheric Research released a study speculating that the Greenland ice sheet could melt faster than previously expected. The study found that 130,000 years ago, when global temperatures last were as warm as they are today, ice melt was such that sea levels eventually rose 20 feet (6.1m) higher than present levels. Meltwater from Greenland and other Arctic sources, the study's authors concluded, contributed 11 feet (3.3m) of the rise. The NCAR press release quotes the study's lead author, Bette Otto-Bliesner, as saying, "Although the focus of our work is polar, the implications are global. These ice sheets have melted before and sea levels rose. The warmth needed isn't that much above present conditions."[11]

The study, however, is based on climate modeling and paleoclimate records. The melt rate and consequent sea level rise are predictions the scientists have made for the future, rather than events already occurring. Some scientists have questioned some of Otto-Bliesner's conclusions. In August

2006 the magazine *Science* published a letter from several climate researchers that stated that though Greenlandic weather and ice should be carefully monitored, "[we should] not conclude at this moment that 'sea-level rise could be faster than widely thought,' as stated by [the study's authors]."[12]

> **The Greenland ice sheet has been shrinking at the rate of twelve cubic miles (50 cu. km) per year.**

One potentially damaging consequence of a rise in sea levels is the impact on the globe's coasts, which are heavily populated. As water rises higher, it will naturally flood beaches and low-lying areas, including those used for farming. In February 2007 the World Bank released the results of a study which estimated that "even a one meter rise would turn at least 56 million people in the developing world into environmental refugees."[13] Those who would be impacted live in over 84 developing countries, with the most people located in Vietnam, parts of Egypt (including the Nile Delta), Mauritania, Suriname, Guyana, French Guiana, Tunisia, the UAE, the Bahamas, and Benin. With a 3.3-foot (1m) rise, the study notes, Vietnam would be particularly hard hit, with 10 percent of the nation's population displaced.

The Ocean Conveyor Belt

Discussion surrounding the effects of increased global temperature on the state of the world's ocean currents exemplifies the difficulty of accurately predicting the consequences of global warming. Ocean currents are linked in a long conveyor belt, which is kept moving by differences in water temperature and salinity.

Cold, salty water like that around the Arctic and in the North Atlantic is heavy compared to warmer, fresher water. This cold, salty, heavier water is always sinking toward the bottom of the ocean. This process, called a thermohaline pump, keeps the currents moving in the ocean conveyor belt. The belt draws warm water from the tropics to the poles, moving heat around the globe in a set pattern of wind and ocean currents—especially the one known as the Gulf Stream. It is this proximity to the warm Gulf Stream that keeps the coast of Western Europe warm even though it is at high latitude.

Because warmer global temperatures increase the amount of freshwater in the ocean, as well as its overall temperature, scientists have predicted the conveyor belt could slow, or even stop, causing warm air to cease flowing over Europe and even cooling the region's climate significantly. The magazine *New Scientist* reported a 30 percent reduction in the conveyor belt in December 2005. In September 2006 the magazine noted, "Climate models . . . [raise] concerns about the flow of deep waters through the North Atlantic—driven by the sinking of dense, salty water at high latitudes. Disrupting this current could cause temperatures to plunge across northern Europe."[14]

> **One potentially damaging consequence of a rise in sea levels is the impact on the globe's coasts, which are heavily populated.**

However, in November 2006, only two months later, the magazine *Science* ran an article with the headline "False Alarm: Atlantic Conveyor Belt Hasn't Slowed Down After All." More accurate measurements had shown the slowing reported in 2005 to be just a natural blip. "There's no sign that the ocean's heat-laden 'conveyor' is slowing," article author Richard Kerr wrote. "Furthermore, researchers are finding that even if global warming were slowing the conveyor and reducing the supply of warmth to high latitudes, it would be decades before the change would be noticeable above the noise."[15]

Furthermore, even if the worst was to happen and the thermohaline pump was to slow or shut down, scientists at the Woods Hole Oceanographic Institution emphasize that the Gulf Stream will never shut off entirely; this is simply not possible. They write, "The winds will continue to blow over the ocean and the Gulf Stream will continue to flow even if the thermohaline circulation slows or shuts down. Its flow may be reduced, or its route slightly redirected, but it will continue to flow."[16]

Melting of Subarctic Permafrost

The public does not tend to pay much attention to places like the Arctic or Siberia, yet the events in those regions are an essential piece of the global warming picture. The land in these areas has a layer of permanently frozen soil called permafrost beneath a layer of topsoil. Increased global temperatures are causing this layer to thaw for the first time in millennia. Some of the effects are obvious: Structures such as roads and houses that were built on the frozen ground are destabilized as the permafrost melts. Craters appear in roads, houses collapse into the ground, and trees list drunkenly in the mushy, unfrozen topsoil in places like northern Canada, Siberia, and Alaska.

> **"The frozen soil contains large amounts of carbon trapped within it."**

Damage to structures that can be rebuilt is only one story. There is a more important, less obvious effect of global warming on the permafrost. The frozen soil contains large amounts of carbon trapped within it. When it melts, that carbon is released into the atmosphere in the form of methane, a greenhouse gas. Jim Hansen of the Goddard Institute for Space Studies states, "[The] release of methane from melting permafrost . . . has probably been responsible for some of the largest warmings in the Earth's history."[17] In his book *An Inconvenient Truth* Al Gore provides a case study of an area in Siberia—over 386,000 square miles (1 million sq. km) of tundra. This particular area is expected to melt by 2050. When it does, it will release 70 billion tons (63.5 billion t) of stored carbon—ten times the amount emitted by humans each year.

Science News reports that, altogether, 500 billion tons (454 billion t) of carbon are still locked away in the permafrost of the world, two-thirds of the total amount now present as carbon dioxide, methane, and other greenhouse gases in the entire atmosphere. Ecologist Kathy Walter of the University of Alaska has called the warming of Siberian permafrost "a climate time bomb."[18]

Siberian Lakes

In addition to melting permafrost, the vast frozen region of Siberia has many lakes that are warming up. These lakes have recently been found to be releasing five times more methane than previously thought: 4.2 million tons (3.8 million t) per year. Walter explains that when the permafrost melts it releases carbon into the lakes. Bacteria then eat this carbon, which has been locked in the permafrost for 20,000 to 40,000 years, and release it in the form of methane.

The melting permafrost has potential benefits as well. The *Economist* magazine points out that huge portions of Russian Siberia could become habitable for humans year-round as the ground softens and temperature rises. Moreover, undiscovered reserves of natural oil and gas in the northern regions of that country would become accessible; scientists estimate that Siberia and the Arctic contain 25 percent of the world's undiscovered oil and gas resources.

Global Warming and Hurricane Strength

There is strong consensus, however, that hurricane strength has been increasing and is set to increase in the as-yet-unknown future. A study by the National Center for Atmospheric Research showed that category 4 and 5 hurricanes have increased by 80 percent in the last 30 years. However, the link to global warming is more tenuous. Many scientists, including those at the National Oceanic and Atmospheric Administration, believe the intensity of the hurricanes is a part of the Atlantic Multidecadal Oscillation, a natural cycle. But others disagree, saying that increased sea surface temperatures offer more energy for already developing hurricanes. In August 2006 James Eisner from Florida State University published a study in *Geophysical Research Letters* proving the first direct link between climate change and hurricane intensity. According to *Astronomy* magazine, the study concluded that the changing climate will make Atlantic hurricanes stronger, and global warming will further increase damage by these storms. It is important to note, however, that this research predicts an increase in the intensity of hurricanes, not the frequency.

These lakes have recently been found to be releasing five times more methane than previously thought: 4.2 million tons (3.8 million t) per year.

Global Warming and Disease

Global warming could bring increased disease to plants, animals, and humans. As Earth warms, certain areas become more hospitable to microbes, insects, and other pests and germ-carrying animals that would otherwise not be able to survive in that climate. Shorter, warmer winters mean longer seasons for pests to live. In 2006 the *Washington Post* reported that West Nile virus, previously unable to survive in the climate of the United States, has killed hundreds since it first appeared in 1999. The link to climate change is supported by researchers from Harvard who reported in a 2005 paper that insects were carrying diseases from warmer climates into cooler ones. This was particularly noticeable in higher elevations, the report noted, as the insects headed farther and farther up mountains with each warming trend.

> **A study by the National Center for Atmospheric Research showed that category 4 and 5 hurricanes have increased by 80 percent in the last 30 years.**

Global Warming's Benefit for Agriculture

Global warming also has a potential beneficial effect on agriculture, especially in the short term: Farmers in areas with significant warming are already finding that they are able to graze their animals longer outdoors as the frost dates stretch later into the fall. The *New York Times* quotes a Welsh farmer as saying, "Every day that the sheep can eat grass instead of us having to carry out cake is a bonus. . . . We should look at the effects of global warming and learn to work with it and use it effectively."[19]

A 2005 report from the National Center for Policy Analysis, a policy research organization, agrees, stating, "[A warmer planet] results in longer growing seasons—more sunshine and rainfall—while summertime high temperatures change little. A warmer planet means milder winters and fewer crop-killing frosts."[20] The report states that since 1950, warming has increased the world's grain production from 700 million tons (635 million t) to 2 billion tons (1.8 billion t) in 2004.

Primary Source Quotes*

What Are the Consequences of Global Warming?

"Starting with the science, the question is: do we know enough to act? And the answer is unequivocally yes. Every year . . . the science on climate change becomes more certain and more disturbing."

—Eileen Claussen, "Climate Change, the State of the Question and the Search for Answers," speech, St. John's University, October 5, 2006.

Claussen is the president of the Pew Center on Global Climate Change.

"Climate change presents very serious global risks and it demands an urgent global response."

—Nicholas Stern, *The Stern Review on the Economics of Climate Change*, October 30, 2006.

Stern is an economist and author of the landmark study *The Stern Review on the Economics of Climate Change.*

Bracketed quotes indicate conflicting positions.

* Editor's Note: While the definition of a primary source can be narrowly or broadly defined, for the purposes of Compact Research, a primary source consists of: 1) results of original research presented by an organization or researcher; 2) eyewitness accounts of events, personal experience, or work experience; 3) first-person editorials offering pundits' opinions; 4) government officials presenting political plans and/or policies; 5) representatives of organizations presenting testimony or policy.

“[Our] planet’s climate can change, tremendously and unpredictably. Beyond that we can conclude . . . that it is *very likely* that significant global warming is coming in our lifetimes. This surely brings a likelihood of harm, widespread and grave.”

—Spencer Weart, “The Discovery of Global Warming: A Personal Note,” American Institute of Physics, March 2006.

Weart, a historian and physicist, is director of the Center for the History of Physics at the American Institute of Physics.

“[Predictions] of the extent of future warming are based on implausible scientific and economic assumptions, and then negative impacts of predicted warming have been vastly exaggerated.”

—Competitive Enterprise Institute, “About Global Warming,” 2007.

The Competitive Enterprise Institute is a nonprofit public policy organization that supports free markets and limited government.

“While no one event is diagnostic of climate change, the relentless pace of unusually severe weather since 2001—prolonged droughts, heat waves of extraordinary intensity, violent windstorms and more frequent ‘100-year’ floods—is descriptive of a changing climate.”

—Paul Epstein and Evan Mills, *Climate Change Futures: Health, Ecological, and Economic Dimensions*, Center for Health and the Global Environment, Harvard Medical School, November 2005.

Epstein is the associate director of Harvard Medical School’s Center for Health and the Global Environment. Mills is an economist, engineer, and policy analyst for the Department of Energy’s Lawrence Berkeley National Laboratory.

“Significant increases in the average temperature of the earth’s surface will endanger human health, increase the intensity of extreme weather events such as storms, floods, and droughts, and damage fragile ecosystems.”

—Union of Concerned Scientists, “Position Papers of UCS: Global Warming,” August 10, 2005.

The Union of Concerned Scientists is a group of scientists and citizens that lobbies for various scientific and environmental policies.

“Many scientists long known for their caution are now saying that warming has reached dire levels, generating feedback loops that will take us perilously close to a point of no return.”

—Kofi Annan, “As Climate Changes, Can We?” *Washington Post*, November 8, 2006.

Annan is the former secretary-general of the United Nations.

“For sea level, a continued rise of about 10 cm/century for many centuries is the best estimate . . . profound long-term impacts on low-lying island communities . . . seem inevitable.”

—Tom Wigley, “The Climate Change Commitment,” *Science*, 2005.

Wigley is a meteorologist and senior scientist at the National Center for Atmospheric Research in Boulder, Colorado.

"Global sea level has risen about an inch and a quarter in the past 10 years . . . most of this rise in sea level is due to expansion of the ocean as it warms up, and maybe 20 to 35% is from melting of glaciers."

—Kevin Trenbarth, "On the Record: Hurricanes and Global Warming," National Center for Atmospheric Research, June 2006.

Trenbarth is a climatologist and head of the Climate Analysis Section of the National Center for Atmospheric Research.

"Of course, we can't prove [Hurricane Katrina] occurred because of climate change, but we can show with increasing confidence that the intensity, frequency and power dissipation of high severity hurricanes is on the rise."

—Jeffrey Sachs, "The Politics of Global Climate Change," Zucker Fellow's Public Lecture, New Haven, Connecticut, February 27, 2006.

Sachs is an economist and the director of the Earth Institute at Columbia University.

"There are a number of different possible tipping points in the climate system. . . . One of these is the melting of glaciers. . . . Second is the thawing of tundra."

—Daniel Schrag, *EurekAlert! Expert Chat: Climate Change*, transcript. 2005. www.eurekalert.org.

Schrag is a professor of earth and planetary sciences at Harvard University.

"The long-term records of the near-surface permafrost temperature, obtained from different parts of the permafrost zone in northern regions, show a significant warming trend during the last 30 years."

—Vladimir Romanovsky, "How Rapidly Is Permafrost Changing and What Are the Impacts of These Changes?" *NOAA: Arctic Theme Page*, 2004.

Romanovsky is a geophysicist and associate professor at the University of Alaska–Fairbanks.

Facts and Illustrations

What Are the Consequences of Global Warming?

- The Natural Resources Defense Council reports that most of the United States has already warmed, in some areas by as much as **4°F** (2.2°C). No state in the lower 48 states experienced below average temperatures in 2002.

- According to the Worldwatch Institute, Arctic snow cover has declined about **10 percent** over the past 30 years.

- According to *New Scientist*, if the Greenlandic ice cap melted it would raise global sea levels by **21.3 feet** (6.5m).

- The Marine Biological Laboratory reports that the increased rainfall and river flow between 1965 and 1995 into the Arctic Ocean was **4,800** cubic miles (20,000 cu. km).

- Global sea levels have been rising at a rate of **0.04 to 0.1 inches** (1mm to 2.5mm) a year, states the Worldwatch Institute.

- A 2004 study by the IPCC found that melting ice accounted for **two-thirds to three-quarters** of sea level rise in recent decades.

- According to the Woods Hole Oceanographic Institution, the thermohaline pump in the Atlantic Ocean shut down about **12,000** years ago for about 1,000 years.

Sea Ice Cover Decline

This graph from NASA's Earth Observatory shows that in 2005, sea ice was at its lowest level since 1979. Sea ice has declined steadily over the last two and a half decades; the graph also shows that in 2002, 2003, and 2004 sea ice was well below the average of 1979–2000.

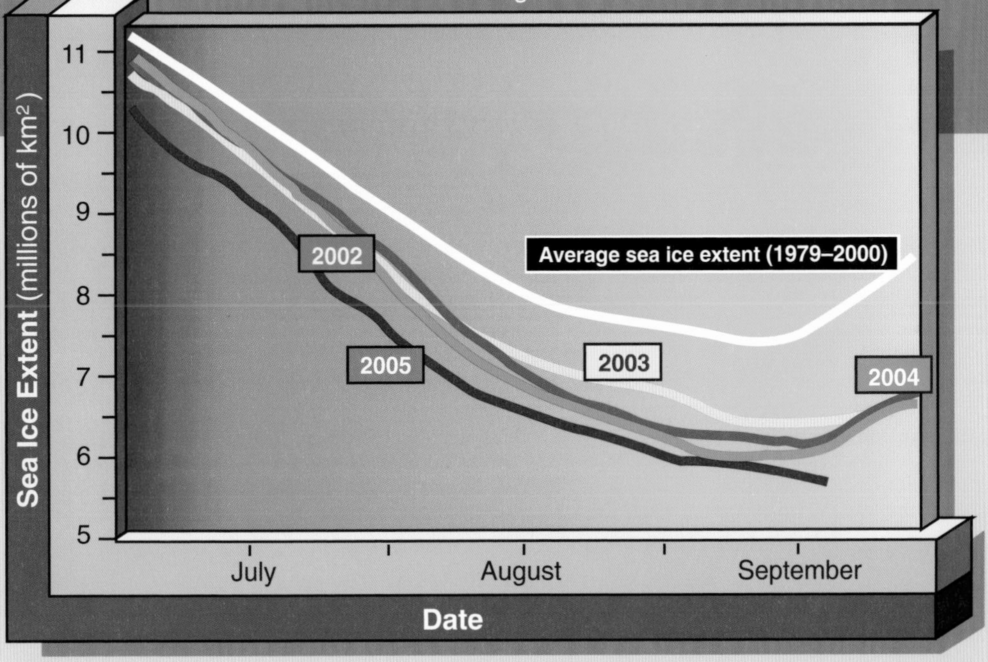

Source: Earth Observatory, NASA, "Continued Sea Ice Decline in 2005," January 8, 2007. http://earthobservatory.nasa.gov.

- According to ecologist Kathy Walter of the University of Alaska, melting Siberian lakes are emitting **4.2 million tons** (3.8 million t) of methane per year.

- Siberia lost **15,444 square miles** (40,000 sq. km) of trees to forest fires in 2003, a record number, states Haiko Balzter of the Unviersity of Leicester.

- According to Ted Schuur at the University of Florida–Gainesville, **551 billion tons** (500 billion t) of carbon is stored in the permafrost of Siberia, two-thirds of the total amount now present in the atmosphere.
- The *Economist* estimates that as much as **25 percent** of the world's undiscovered oil and gas reserves are located in Russia and the Arctic.

Rise in Ocean Temperature

This graph shows the increase in world ocean temperatures from 1955 to 2003. The temperature is measured at three different depths; all follow a similar upward path.

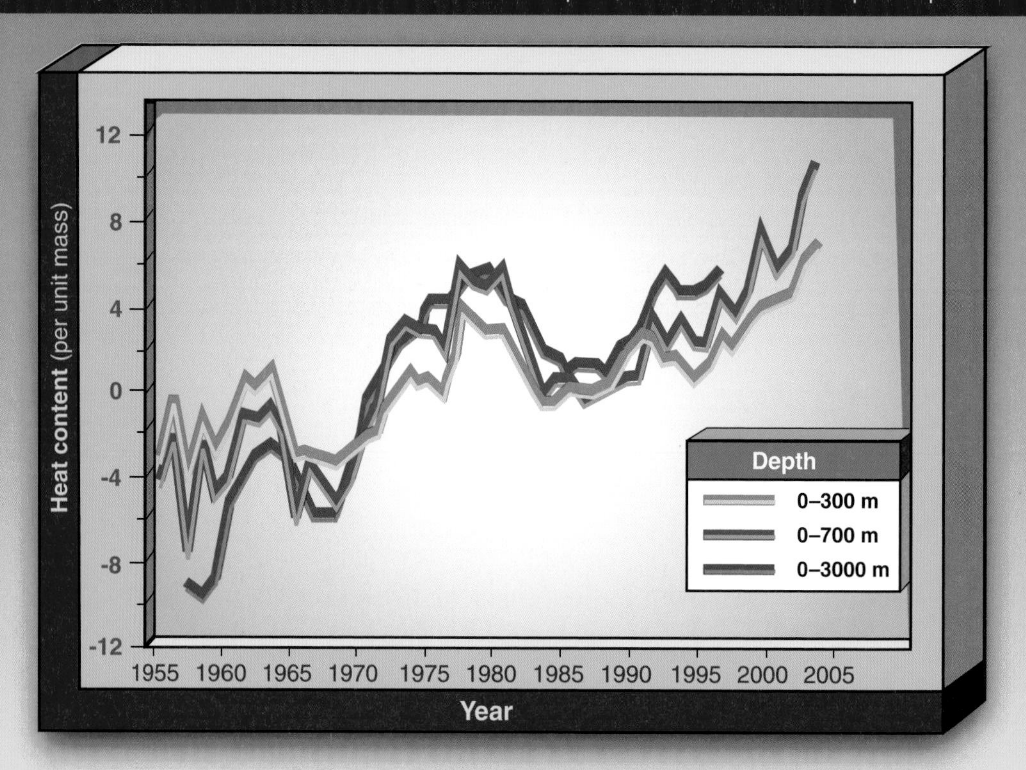

Source: Environmental Defense Fund, "Global Warming, Ocean Cooling?" October 11, 2006. www.environmentaldefense.org.

Thermohaline Circulation and the Ocean Conveyor Belt

This illustration demonstrates the process of thermohaline circulation, which keeps the conveyor belt of the ocean moving and warm air circulating over Europe. If global warming continues, some scientists theorize that it could cause the conveyor belt to halt, drastically cooling the continent.

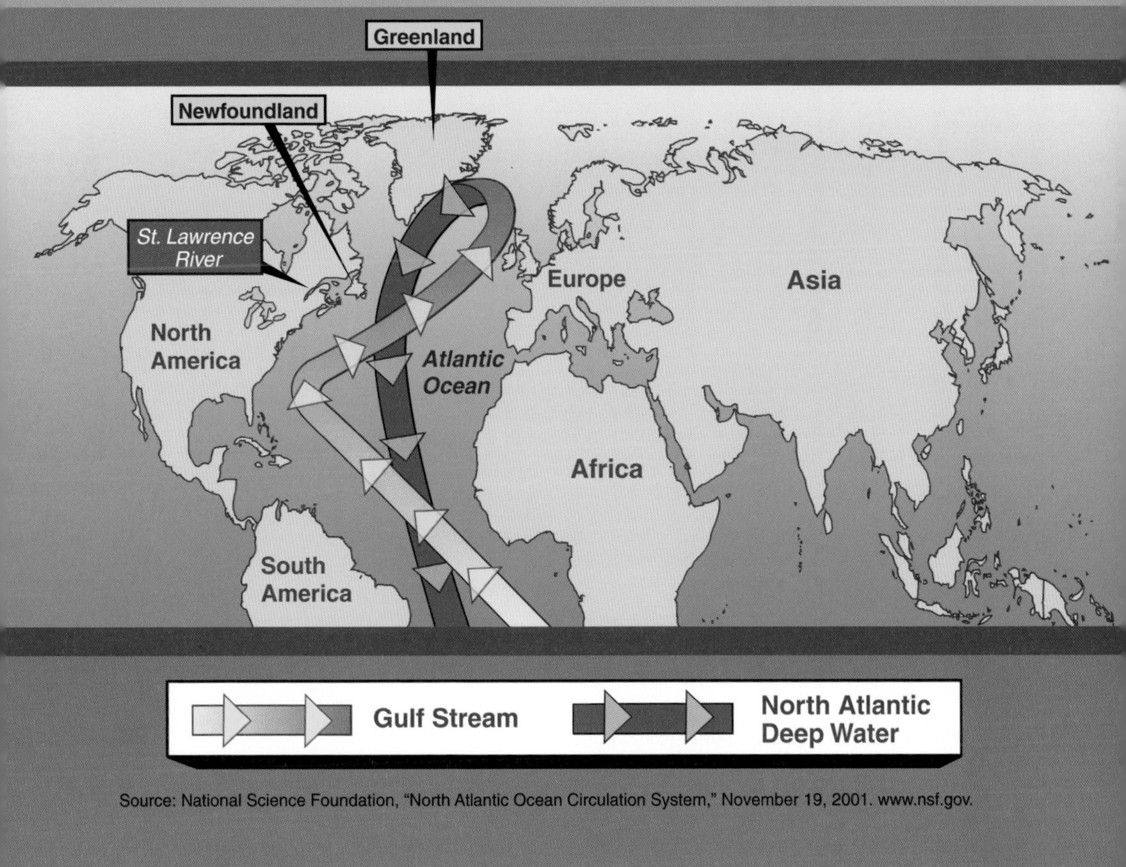

Source: National Science Foundation, "North Atlantic Ocean Circulation System," November 19, 2001. www.nsf.gov.

- Mosquitoes carrying dengue fever viruses were previously limited to elevations of **3,300 feet** (1,006m) but recently appeared at **7,200 feet** (2,195m) in the Andes Mountains of Colombia, states the Natural Resources Defense Council.

Economic Losses of Extreme Weather Events

This data, provided by the Swiss reinsurance company Munich Re, demonstrates the expense catastrophic natural events such as flooding, hurricanes, and earthquakes can cause. Flooding in Europe, for instance, cost $60 billion in 2002. As global warming continues, scientists predict that these events will become more common.

Year	Number of Events	Number of Victims	Total Losses (millions of U.S. dollars)	Insured Losses (millions of U.S. dollars)	Major Events
2000	850	10,300	$38,000	$9,400	Floods (UK), Typhoon (Samoi)
2001	720	25,000	$40,000	$11,800	Tropical storm Allison, Hailstorm (USA)
2002	700	11,000	$60,000	$14,000	Floods (Europe)
2003	700	89,000	$65,000	$16,000	Heat wave (Europe), Earthquake (Bam/Iran)
2004	650	225,000	$150,000	$47,000	Hurricanes (Atlantic), Typhoons (Japan), Tsunami
2005	670	101,000	$219,000	$99,000	Hurricanes (Atlantic), Earthquake (Pakistan)
2006	850	18,000	$45,000	$15,000	Earthquake (Yogyakarta/Indonesia)

Source: Geo Risks Research, Munich Re, "Natural Catastrophes 2006." December 2006. www.munichre.com.

- In 2006, scientists in Spain confirmed the first evidence of a **species decline** directly linked to climate change, according to *New Scientist*.

- In 2006, *Foreign Policy* magazine reported that with a **5 degree increase** in global temperatures countries such as Canada and Russia could experience over a **200 percent** increase in tourism.

- The East Coast of the United States has experienced **fewer effects** of climate change than the rest of the world. The most obvious signs of global warming have occurred in the **Arctic** and northern regions such as Siberia and Greenland.

What Are the Controversies Surrounding Global Warming?

"Consensus as strong as the one that has developed around this topic is rare in science."

—Donald Kennedy, "An Unfortunate U-Turn on Carbon."

"[The] work of science has nothing whatever to do with consensus. Consensus is the business of politics."

—Michael Crichton, "Aliens Cause Global Warming," Lecture.

Controversy over Cause of Global Warming

Few in the public or the scientific community dispute that Earth is indeed getting warmer, but scientists and policy makers are divided on whether the warming is human-induced, a result of natural events, or some combination of the two. One of the best-known climate change events was that which caused the extinction of the dinosaurs during the Cretaceous period 65 million years ago. Some scientists believe massive volcanic eruptions spewed huge amounts of dust, carbon dioxide, and sulfur dioxide into the atmosphere—enough to change global temperatures. Others argue that the climate change was due to the impact of a massive meteor or asteroid. Either way, there is little doubt that climate change can be induced entirely by natural causes. The controversy centers around just how much, if any, of today's warming is due to natural events.

Natural Changes

In August 2006 two scientists from the University of Southern California published a paper in the journal *Environmental Geology* arguing that their research shows that the climate changes Earth is experiencing are due to natural rather than human-induced causes. They claim that solar radiation, outgassing from volcanoes and other sources, and microbial activities are contributing the vast majority of the changes to the atmosphere. The authors write, "Human-induced climatic changes are negligible." Furthermore, they state in their conclusion, "[Any] attempts to mitigate undesirable climatic changes using restrictive regulations are condemned to failure, because the global natural forces are at least 4–5 orders of magnitude greater than available human controls. In addition, application of these controls will lead to catastrophic economic consequences."[21]

> **"There is little doubt that climate change can be induced entirely by natural causes."**

Others, such as science writer Eugene Linden, speculate that Earth is indeed in the midst of a period of rapid natural change but that this change is exacerbated by, though perhaps not entirely the result of, modern civilization. Climate change has wiped out civilizations in the past, Linden writes, but still, "no modern industrial society has been tested by the protracted climate chaos that destroyed [past civilizations]. Nor has the global system of markets and food distribution, which has largely eliminated death by starvation in Africa."[22]

Controversy Surrounding Global Warming Policy

The idea that humans are at least partially responsible for global warming and climate change has vocal opponents in both Congress and the executive administration. One of these opponents is the president's top environmental adviser, James Connaughton, the chairman of the White House Council on Environmental Quality. *American Prospect* magazine reported that Connaughton told the BBC, "We are still working on the issue of causation, the extent to which humans are a factor."[23]

A particularly vehement opponent of the harms and reasons behind global warming is Senator James Inhofe of Oklahoma, former chairman of the Senate Committee on the Environment and Public Works. Inhofe has vigorously denied the current scientific consensus on global warming, memorably calling the issue "the greatest hoax ever perpetrated on the American people." He received a great deal of negative media attention for this statement and defended himself by saying,

"The idea that humans are at least partially responsible for global warming and climate change has vocal opponents in both Congress and the executive administration."

> I have insisted all along that the climate change debate should be based on fundamental principles of science. . . . Ultimately, I hope, it will be decided by hard facts and data—and by serious scientists committed to the principles of sound science. Instead of censoring skeptical viewpoints, as my alarmist friends favor, these scientists must be heard and I will do my part to make sure that they are heard.[24]

There are claims on both sides that not all views are being heard on global warming. Thomas J. Crowley, a climate scientist at Duke University, wrote in an article titled "Are Claims of Global Warming Being Suppressed?": "[Over] the last few years, I have heard many rumors that climate science relevant to the global warming discussion is being suppressed by the [George W.] Bush Administration."[25]

Benefits vs. Harm

Many organizations, such as the Union of Concerned Scientists, the American Association for the Advancement of Science, and the Intergovernmental Panel on Climate Change, have issued cautionary statements warning of the potential harm of continued global warming. The AAAS, for instance, released a statement in February 2007 that reads in part, "The scientific evidence is clear: global climate change caused by human activities . . . is a growing threat to society. . . . The time to control greenhouse gas emissions is now." [26]

"There are claims on both sides that not all views are being heard on global warming."

Other organizations disagree, however, and argue that, on the contrary, global warming can benefit the planet, whether brought on by humans or nature. The *World Climate Report*, an online newsletter/blog published by the Greening Earth Society, a group created by the Western Fuels Association, is one such publication offering this point of view. For instance, in an article titled "Global Forests Love Global Warming" the report argues that higher carbon dioxide levels are beneficial to forests rather than harmful. Another article records the expansion of a species of lizard in an area experiencing warming as a result of elevated carbon dioxide levels. The report offers these articles, it says, to counterbalance reporting that global warming is harmful to Earth.

Warmer Winters

A similar argument is made by Myron Ebell, director of energy and global warming policy at the Competitive Enterprise Institute, a pro-market, free enterprise public policy group. In a 2006 article in *Forbes* magazine, Ebell writes that when the prospect of warmer winters is taken into account, global warming seems less harmful than helpful. "At the risk of committing heresy," Ebell writes,

> I'd like to suggest that the debate about climate change include, for once, a fair assessment of the benefits alongside the declamations of harm. For example, cold winter storms kill a lot of people. . . . So modest climatic improvement would be to have fewer and less severe big winter storms. Amazingly, that's exactly what we should get if global warming theory turns out to be true.[27]

Furthermore, Tom Bethell of *American Spectator* writes that policies such as the Kyoto Protocol, meant to benefit the planet, could actually wind up harming the United States economy. Bethell notes, "Under the Kyoto Protocol, U.S. emissions would have to be cut so much that economic depression would have been the only certain outcome. We were expected to reduce energy use by about 35 percent within ten years, which might have meant eliminating one-third of all cars." [28]

Controversy Surrounding the *Stern Review*

The economics of climate change have also been a source of controversy. In 2006 the British government commissioned and released the *Stern Review*, a massive document prepared by Nicholas Stern, former chief economist of the World Bank, that discussed the economic impact of global warming. The report was well received and well respected but some have accused Stern of unrealistic fearmongering. *Newsweek*'s Robert Samuelson in particular has taken issue with the report. In general, Stern predicts massive economic costs directly related to global warming–induced flooding, drought, and disease. Yet switching to alternative fuels and thus stabilizing carbon emissions could cost as little as 1 percent of global output, according to the report. Samuelson takes issue with these conclusions, stating, "The report is a masterpiece of misleading public relations. . . . Stern's headlined conclusions are intellectual fictions. They're essential fabrications to justify an aggressive anti-global-warming policy. The danger of that is we'd end up with the worst of both worlds: a program that harms the economy without much cutting of greenhouse gases."[29]

> **"Other organizations . . . argue that, on the contrary, global warming can benefit the planet, whether brought on by humans or nature."**

Samuelson argues that Stern is not facing the serious economic realities of mitigating global warming. There is a reason, he writes, that the public has not demanded the federal government take comprehensive action on climate change—as a whole, society has to *want* to. Samuelson offers three reasons that society is not yet at that point; reasons that the technology needed to cut greenhouse gas emissions is not "economically

> **The economics of climate change have also been a source of controversy.**

acceptable"[30] because coal is cheap, alternative energy is not—at least at this point in time. Politicians would have to make unpopular choices that would inconvenience their constituents now in order to gain benefits that they would never live to see. This is highly unlikely. The third reason Samuelson offers is that developing countries refuse to cut their emissions because it might hurt their growth—and the developed world is not requiring them to.

A Shift in Values

Ending human reliance on greenhouse gas–producing substances would mean in essence an overhaul of the way we think about and use energy; a sea change in how we live and function. To illustrate the profundity of this change, scientist Robert Socolow has compared the current state of energy production to that of child labor in the United States. Child labor was an effective system that allowed Americans access to cheap goods, Socolow states, just as coal and natural gas provide cheap energy. Yet at some point in history, society as a whole decided that child labor was morally wrong. Society abolished child labor and as a result, the price of goods went up. But morally, it was worth it. Socolow also mentions that decisions made by American society to address the needs of the disabled and mitigate air pollution were based on morality rather than economics.

Primary Source Quotes*

What Are the Controversies Surrounding Global Warming?

"Human activities are increasingly altering the Earth's climate. . . . Scientific evidence strongly indicates that natural influences cannot explain the rapid increase in global near-surface temperatures observed during the second half of the 20th century."

—American Geophysical Union, "Position Statement: Human Impacts on Climate," December 2003.

The American Geophysical Union is an association of geophysicists that works to advance the public's understanding of science and the environment.

"'Climate change is real' is a meaningless phrase used repeatedly by activists to convince the public that a climate catastrophe is looming and humanity is the cause. Neither of these fears is justified."

—James Inhofe, "Hot & Cold Media Spin Cycle," speech, U.S. Senate floor, September 25, 2006.

Inhofe is a Republican U.S. senator from Oklahoma and the former chairman of the Senate Environment and Public Works Committee.

Bracketed quotes indicate conflicting positions.

* Editor's Note: While the definition of a primary source can be narrowly or broadly defined, for the purposes of Compact Research, a primary source consists of: 1) results of original research presented by an organization or researcher; 2) eyewitness accounts of events, personal experience, or work experience; 3) first-person editorials offering pundits' opinions; 4) government officials presenting political plans and/or policies; 5) representatives of organizations presenting testimony or policy.

"Consensus as strong as the one that has developed around this topic [global warming] is rare in science."

—Donald Kennedy, "An Unfortunate U-Turn on Carbon," *Science*, March 2001.

Kennedy is the editor in chief of the journal *Science*.

"We've lived long enough to see more than one 'consensus' blown apart within a few years. . . . Perhaps it's useful to have a few folks outside the 'consensus' asking questions."

—*Wall Street Journal*, "Global Warming Gag Order," December 4, 2006.

The *Wall Street Journal* is a leading financial and business newspaper.

"Climate change denial has been so effective because the 'denial community' has mischaracterized the necessarily guarded language of serious scientific dialogue as vagueness and uncertainty."

—Olympia Snowe and John D. Rockefeller, letter to Rex W. Tillerson, chairman and chief executive officer, ExxonMobil Corporation, October 27, 2006.

Snowe is a Republican U.S. senator from Maine. Rockefeller is a Democratic U.S. senator from West Virginia.

"In the United States of America, unfortunately we still live in a bubble of unreality [with regard to climate change]. . . . Nobody is interested in solutions if they don't think there's a problem."

—Al Gore, in David Roberts, "Al Revere: An Interview with Accidental Movie Star Al Gore," *Grist.org*, May 2006.

Gore is the former vice president of the United States and author of the book and documentary about climate change, *An Inconvenient Truth*.

“Individual scientific statements and papers . . . can be exploited . . . and can leave the impression that the scientific community is sharply divided on issues where there is, in reality, a strong scientific consensus.”

—*Bulletin of the American Meteorological Society*, “Climate Change Research Issues for the Atmospheric and Related Sciences,” February 9, 2003.

The American Meteorological Society is a scientific organization of meteorologists and climate scientists.

“[Global] warming would likely . . . actually *reduce* the number of lives lost to extreme thermal conditions, as many more people die from unseasonably cold temperatures than from excessive warmth.”

—Center for the Study of Carbon Dioxide and Global Change, “Enhanced or Impaired? Human Health in a CO_2-Enriched Warmer World,” November 3, 2003.

The Center for the Study of Carbon Dioxide and Global Change is an organization that studies the effects of increased carbon dioxide in the atmosphere and the potential beneficial effects of increased CO_2.

“[When] it comes to an issue like climate change . . . the government ought to be bending over backward to make sure that its scientists are able to discuss their work and what it means.”

—Sherwood Boehlert, letter to NASA administrator Michael Griffin, January 30, 2006.

Boehlert is a U.S. congressman from New York and the former chairman of the House Science Committee.

“There are almost always divergent [scientific] views and weighting these is frequently very difficult. We ask that Congress respect this time-tested process of scientific quality control.”

—John Orcutt and Walter Lyons, letter to Joe Barton, chairman, Committee on Energy and Commerce, August 8, 2005.

Orcutt is the president of the American Geophysical Union. Lyons is the president of the American Meteorological Society.

“I believe the overarching problem is that the [Bush] administration . . . does not want and has acted to impede forthright communication of the state of climate change and its implications for policy.”

—Rick Piltz, “On Issues of Concern About the Governance and Direction of the Climate Change Science Program,” letter to U.S. Climate Change Science Program Agency Principals, June 1, 2005.

Piltz is a former senior associate with Climate Change Science Program and founder of Climate Science Watch, a government watchdog group.

“Action to prevent runaway global warming may prove cheap, practical, effective, and totally consistent with economic growth.”

—Gregg Easterbrook, “Some Convenient Truths,” *Atlantic Online*, September 2006. www.theatlantic.com.

Easterbrook is a contributing editor at the *Atlantic Monthly* magazine and a fellow at the Brookings Institution, a think tank.

“The *Stern Review* shows us, with utmost clarity, while allowing fully for all the uncertainties what global warming is going to mean; and what can and should be done to reduce it.”

—James Mirrlees, “Responses to the *Stern Review*,” *Stern Review* Index Page, October 30, 2006. www.hm-treasury.gov.co.uk.

Mirrlees won the Nobel Prize in 1996 for economics. He is professor emeritus of political economy at the University of Cambridge in England.

Facts and Illustrations

What Are the Controversies Surrounding Global Warming?

- Over the last **500 million years** 5 major extinctions have occurred, all of which killed over 25 percent of all life, and 3 of which killed over **70 percent** of all species. All 5 extinctions were triggered by natural causes, according to *New Scientist.*

- According to science writer Eugene Linden, rapid climate change is capable of obliterating civilizations within the space of a few generations.

- A July 2006 poll by the Pew Research Center found that **70 percent** of Americans surveyed believe there is solid evidence that Earth is warming, but only 41 percent believe it was due to human activity.

- According to a 2006 survey by the Massachusetts Institute of Technology, Americans now say **climate change** is the country's most pressing environmental problem, up from number 6 out of 10 concerns in 2003.

- The National Center for Policy Analysis has stated that since 1950, additional carbon dioxide in the environment has helped global grain production increase from **700 million tons** (635 million t) to 2 billion tons (1.8 billion t) in 2004.

- According to Gregory Jones, a professor of geography at Southern Oregon University, increased temperatures since 1950 have improved the quality of certain **wine grapes** in the United States.

Scientific and Public Opinion Diverge

In the following graph, author Eugene Linden charts the vast difference between the view of science on climate change and the view of the public from 1988 to 2005. The scientific community's view has gone steadily from indifference to alarm with a general consensus. In contrast, the public has mostly been indifferent over time, except for a brief spike toward alarm in the late 1980s.

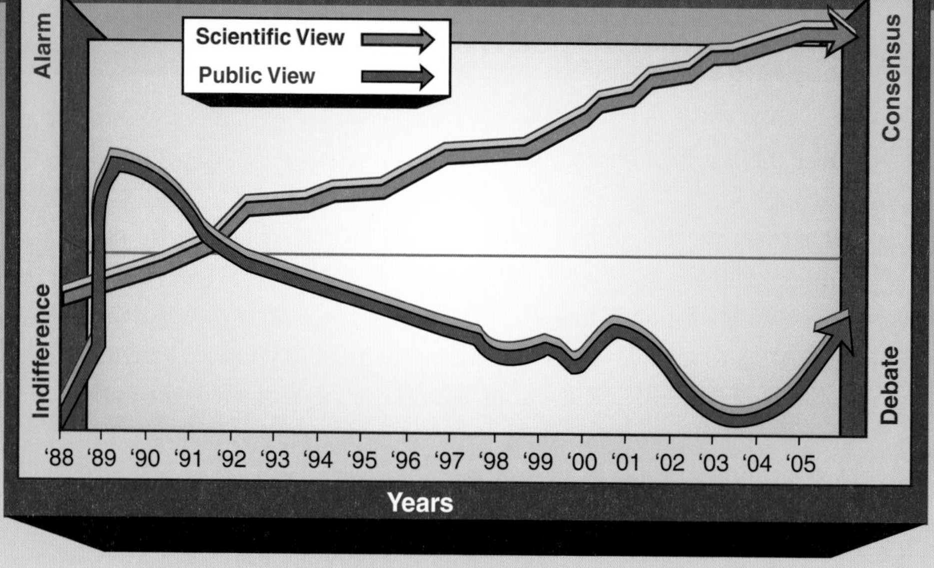

Source: Eugene Linden, *The Winds of Change*, 2006.

- Richard Muller, a physicist at the University of California at Berkeley, has suggested an unusual cosmic source for cooling cycles that occur roughly every **100,000 years**.

- *The New Oxford American Dictionary* recently proclaimed **"carbon neutral"** as its Word of the Year for 2006.

- The volume of metric tons of carbon traded on the voluntary market **doubled** in 2006 over 2005, according to the *Christian Science Monitor*.

- In a poll of 92 companies by the Conference Board, a nonprofit business research organization, **75 percent** were considering their carbon footprint.

- The ***Stern Review*** is one of the first major government-sponsored reports on global warming conducted by an economist and not an atmospheric scientist.

Americans Do Not Agree on Global Warming

According to a Pew Research Center poll, Americans generally agree that global warming is occurring but they do not agree on why this is happening. Seventy percent of all people surveyed agreed there is solid evidence that the earth is warming, but only 41 percent said it is due to human activity.

Source: Pew Research Center, "Little Consensus on Global Warming," July, 12, 2006.

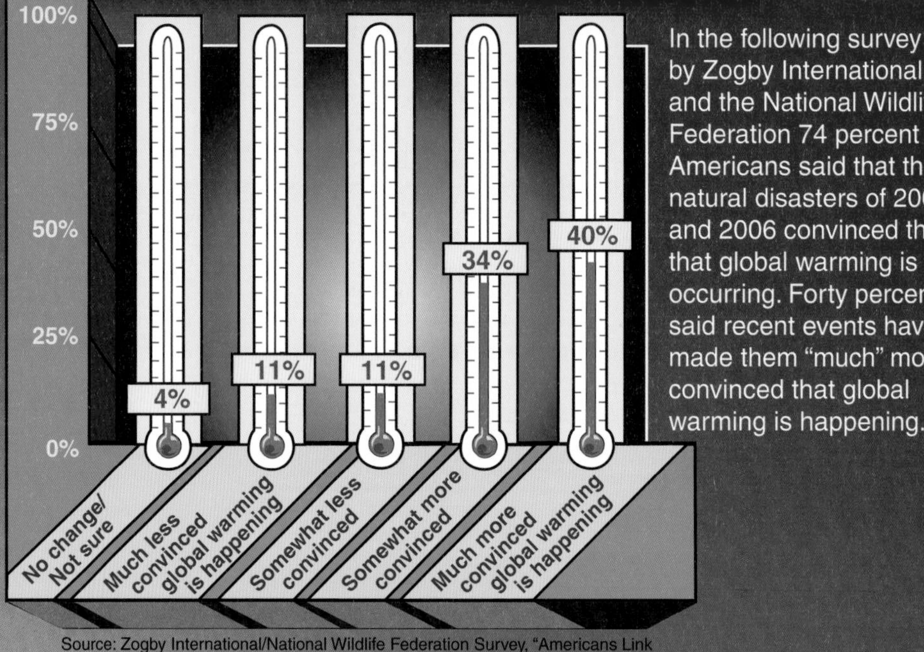

Source: Zogby International/National Wildlife Federation Survey, "Americans Link Hurricane Katrina and Heat Wave to Global Warming," August 21,2006. www.zogby.com.

- The cost of action to reduce greenhouse gas emissions to avoid the worst impacts of climate change can be limited to around **1 percent** of global GDP each year, according to the *Stern Review on the Economics of Climate Change.*

- In a 2005 study by Dr. Naomi Oreske of the University of California-San Diego, a random sampling of **928** peer-reviewed journal articles on global warming from the past 10 years, revealed that **100 percent** agreed with the view that humans affect global warming.

Cost of Alternative Energy Sources

The *Economist* magazine reported in September 2006 that the cost of replacing current energy sources with emissions-reducing technology would be more expensive in every area. Some technologies however, end up more cost-effective over a longer period. For instance, the cost of grid electricity is 10 cents per kilowatt-hour. A replacement by a non-greenhouse-gas-producing photovoltaic system would be 15 cents over the short term, but would drop to 8 cents over the long term.

Illustrative costs of emissions-reducing technology relative to a marker*

Technology	Marker*	Cost/unit	Cost of Marker	Cost of substitute short-term	Cost of substitute long-term
Nuclear	Natural gas combined cycle power plant	US cents/Kwh	3.5–4	6	5
Electricity from fossil fuels and carbon capture and storage	Natural gas combined cycle power plant	US cents/Kwh	3.5–4	5	6
Wind	Natural gas combined cycle power plant	US cents/Kwh	3.5–4	5	6
Photovoltaic	Grid electricity	US cents/Kwh	10	15	8
Biofuels	Petrol (gasoline)	$1/gigajoule	12	15	15

* a current technology that could be displaced by new technology

Source: *Economist*, "Dismal Calculations," September 9, 2006.

What Are the Solutions for Global Warming?

> "I say the debate is over. We know the science. We see the threat. And we know the time for action is now."
>
> —Arnold Schwarzenegger, "Governor's Remarks at World Environment Day Conference."

> "I think [the carbon-emissions reduction strategies have] been relatively simple-minded so far and have failed to take into account . . . unintended consequences."
>
> —James Connaughton, "CEQ's Connaughton Discusses Admin's Current and Future Clean Energy Policies."

Global Political Solutions

Today, the Kyoto Protocol covers about 60 percent of all greenhouse gas emissions. However, the agreement's effectiveness is limited because the United States has refused to ratify it, as has Australia—despite being two of the world's largest producers of greenhouse gases.

If America signed up it would be required to limit its annual carbon emissions to 7 percent below 1990 levels. This requirement does not set the country apart from its peers. The European Union is supposed to produce carbon at 8 percent below 1990 levels and Japan 6 percent. The countries can meet their goals partly by buying and selling credits and also by investing in clean energy projects in developing nations. Kyoto, however, is not a permanent solution: It only provides protocols through 2012.

China, India, and the Developing World

The efforts of the United Kingdom and others are admirable, yet will be in vain if emissions in developing countries are not curbed—particularly in China and India. Neither of these countries is included in the Kyoto framework because of their status as developing nations. Yet both are growing at an explosive rate and rapidly industrializing. Their energy needs have risen in accordance with their population growth. Science writer Elizabeth Kolbert states in her book *Field Notes from a Catastrophe* that China will double its need for electricity by 2050. To meet its increasing energy demand, the nation is building the sort of outdated coal-fired power plants that the United States stopped constructing years ago. Lifetime emissions only from China's *planned* coal plants will be 25 billion tons (22.7 billion t) of carbon. Tony Blair pointed out in November 2006 that even if Britain were to completely end all of its greenhouse gas output immediately, China's emissions alone would negate the gains in only two years.

"Many government officials have made the point that Kyoto will be useless unless it includes India and China."

Many government officials have made the point that Kyoto will be useless unless it includes India and China. The governments of these countries object to Kyoto, citing the strain the regulations would place on their economies. Few dispute that curbing carbon emissions will place more of a burden on the already fragile developing nations. However, economist Nicholas Stern and others argue that if global warming continues, rich countries as well as poor will begin to suffer economically.

In addition to being poor, the countries of Africa and India are also particularly vulnerable to the effects of global warming because of their naturally hot climates and dependency on agriculture. Crops can be easily destroyed by storms, floods, fire, or insects—all problems that could occur with potentially increasing frequency over the next decades.

Understanding Stabilization Wedges

In order to better understand the monumental effort that will be required to slow, halt, or possibly even reverse the human-induced aspects of global

> **Without waiting for the government to act, the [insurance] industry as a whole has raised its premiums in hurricane and flood prone areas.**

warming, Princeton University climate scientist Robert Socolow has developed a system of what he calls "wedges." Each "wedge" represents a step that would prevent a certain amount of emissions per year from being emitted. Many of the wedges are different types of alternative energy: solar, wind, nuclear, and a technology called "carbon capture and storage," in which the carbon emitted by power plants is caught and injected underground. Some wedges are the use of fuel-efficient vehicles, reduced use of vehicles, and the construction of fuel-efficient buildings. Others are forest restoration projects.

The wedge theory assumes that countries or government bodies would use several different wedges at one time in order to reach the desired carbon emissions level. In order to avoid a doubling of preindustrial levels of carbon dioxide, for instance, seven wedges would be needed, each one mitigating one gigaton of carbon per year.

Energy-Based Solutions

While seeking alternative sources of energy is important in mitigating climate change, many believe preparing for the inevitable is also important. John Holdren, president of the American Association for the Advancement of Science and Alan I. Leshner, publisher of *Science*, write, "[Adaptations] will be essential because the climate is already changing and will change more before measures to reduce or counter society's emissions of heat-trapping gases can take hold."[31] The Dutch, living in a country surrounded by water and a complex system of dikes and dams, have already taken this approach. They have built floating houses, assuming that as sea levels rise, much of the land will be underwater one day in the not-too-distant future.

The insurance industry also has accepted that some impact from global warming will be inevitable. Without waiting for the government to act, the industry as a whole has raised its premiums in hurricane and flood prone areas. Insured losses during 2005 alone have already cost the industry $57 billion. Marsh Insurance said in a 2006 memo to clients that "[climate] change . . . is one of the most significant emerging risks facing the world today, presenting tremendous challenges to the environment, to the world economy, and to individual businesses . . . [who] if they haven't already, must begin to account for it in the strategic and operation planning."[32]

The United States

The U.S. government, as of the time of this writing, had no official policy on climate change nor had it imposed mandatory standards on American industry with regard to carbon emissions. Bills that seek to impose caps on carbon dioxide have been introduced in Congress but generally have been opposed by the White House. James Connaughton, the president's chief environmental adviser, has spoken out against what he calls "simple-minded" global warming legislation that could harm the American economy. Instead, Connaughton told the U.S. Energy Association in November 2006 that the administration is focusing its energy priorities on expanding offshore oil drilling.

> **In August 2006 California became the first state in the nation to impose mandatory greenhouse gas emissions caps.**

Despite the lack of support from the White House, both Congress and trade groups are beginning to take steps of their own to curb emissions. In September 2006 Jeffrey E. Sterba, the CEO of a major Texas utility company and future head of a Washington utility trade group, told *Business Week* magazine that "the time has come to do something on climate change, and it is better to act sooner rather than later."[33] On November 29, 2006, the U.S. Supreme Court heard arguments in a case brought by Massachusetts and other states against the federal Environmental Protection Agency. The states argued that the agency has failed to regulate greenhouse gases as it is required to do under the Clean Air Act. A decision was pending at the time of this writing.

State and Local Level Solutions

Frustrated by the federal government's refusal to take action on climate change, some states have also decided to act on their own. In August

2006 California became the first state in the nation to impose mandatory greenhouse gas emissions caps on businesses under a bill called the Global Warming Solutions Act, which was supported by Governor Arnold Schwarzenegger. These are the first caps ever imposed in the United States. The bill requires California to cut greenhouse gas emissions 25 percent to 1990 levels by 2020. In addition, 20 percent of California's energy has to come from sustainable sources by 2010. By 2016 emissions from cars have to be cut by 30 percent.

The California legislature expects to reach these goals with a multiheaded approach: a combination of cleaner transportation, clean power plants, green building, and alternative fuels. It is also working on smart-growth plans for all new development. California business has supported the bill, even though some companies will surely move elsewhere. Many companies worked with the legislature to implement business-friendly applications, like a cap-and-trade system: Companies that cannot meet the new standards can purchase credits from companies that are under the limit.

> **In April 2006 General Electric, Shell, Exelon, and Duke Energy all agreed that they would accept mandatory caps on greenhouse gas emissions.**

In an August 2006 speech California senator Dianne Feinstein reiterated the state's commitment to combating climate change and the benefits it expects to receive from doing so:

> With every challenge comes a new opportunity, and California is well-positioned to take advantage of a new low-carbon economy. The State has already begun to reap the economic benefits of cleaner, greener, and more efficient technologies and standards. . . . Working together, I believe we can reduce our emissions sufficiently to stabilize the Earth's climate, to minimize warming and slow global temperature increases . . . to avoid catastrophic climate change.[34]

The private sector also has begun exploring alternative energy sources, suspecting perhaps that restrictions on fossil fuels will soon be in place, as the *Economist* magazine speculates. In April 2006 General Electric, Shell, Exelon, and Duke Energy all agreed that they would accept mandatory caps on greenhouse gas emissions, should the federal government put them in place. The chemical producer DuPont has taken the idea a step further. Since 1990 the company has voluntarily reduced its own emissions by 72 percent and promises another 15 percent cut by 2015. The company has also begun making a new type of plastic that comes from corn, a renewable energy source, and requires 40 percent less energy to produce than traditional plastic.

General Electric has also recently made a public commitment to carbon-emissions reduction, pledging to double its investment in renewable energy technology. The decision, CEO Jeffrey Immelt said, was based on sound business practice—the company expects to double its profit from the new green technology by 2010, according to *U.S. News and World Report.*

Some of these voluntary measures may be due in some part to a new wave of civil litigation surrounding climate change. In a move that is reminiscent of the tobacco lawsuits of the 1990s, some people have begun to sue companies such as Exxon-Mobil and General Motors for damages to their homes caused by Hurricane Katrina. These lawsuits are not necessarily for the purpose of actually collecting damages, *Business Week* magazine points out, but to pressure industry and government to issue mandatory greenhouse gas caps.

Private citizens can take matters into their own hands in a less litigious way by living and traveling "carbon-neutral." Travelers, for instance, can go to a number of different Web sites prior to a trip, calculate how much carbon dioxide their transportation for the trip will add to the atmosphere, and then buy an "offset"—donate money to alternative energy projects or groups that work to end greenhouse gas pollution. Other sites calculate a person's "carbon footprint"—they take into account how much the person recycles, what sort of car he or she drives, and so on, then calculate the amount to be donated as an offset.

"Private citizens can take matters into their own hands . . . by living and traveling 'carbon-neutral.'"

While the general consensus on the best way to control global warming is to curb emissions that spew greenhouse gases into the atmosphere, some scientists have conceived of ways to cool Earth with physical devices—"geoengineering." Astronomer Roger Angel of the University of Arizona has studied the idea of hanging a sort of sunshade in space between Earth and the sun, using gravity to keep it in place. Another idea involves spraying pollution into the air to block sunlight with reflective particles. Yet another has vessels skimming the ocean's surface, spraying saltwater into clouds to increase their reflectivity. The majority of these ideas are unlikely to be implemented. If some of the less radical ideas are implemented, they will not be long-term solutions but holding mechanisms until alternative energy sources are ready for worldwide usage. Generally, both scientists and economists agree that the best way to end global warming is to reduce the activities that cause it rather than spend more money trying to block the effects.

"Some are convinced that increasing organic farming can help reduce global warming."

Agriculture is a frequently overlooked contributor to global warming. Fertilizers, farm equipment, and liquid manure sitting in tanks all release greenhouse gases. Consequently, some are convinced that increasing organic farming can help reduce global warming. In September 2006 the Soil Association, a group of British organic farmers, reported that farming made up at least 8 percent of greenhouse gas emissions and probably far more, since factors such as the trucking of the goods to market and emissions from the making of fertilizer were not included in that number, according to the magazine *Farmers Weekly*. Conventional farming releases about thirteen tons (11.8t) of carbon per year, but organic farming actually *absorbed* carbon, the association reported. As well, organic farming used less energy from start to finish in the production of its goods.

The Economics of the Solutions

Economists have run estimates on the difference between the costs of living with global warming and the costs of mitigating global warming and have agreed that it will be cheaper by far to mitigate than live with.

According to Robert Mendelson, journalist for the *Economist* magazine, three factors must be considered: the extent to which the demand for energy can be reduced by inexpensive measures, the rate at which renewable technology becomes affordable, and the speed at which emissions are reduced. Mendelson writes that according to the International Energy Association, by 2050 inexpensive measures will be capable of reducing emissions to 2000 levels and that the cost of some renewable technology has already dropped to "affordable" levels. Reducing emissions slowly will be more cost-effective; rather than replacing devices immediately, technology can be used to the end of its lifespan and then replaced with greener, more efficient machines and devices.

Solutions are possible. Humboldt State University professor John M. Meyer points out that reducing global warming will benefit all:

> A serious push to retrofit buildings . . . to foster in-fill development rather than sprawl, or to restore wetlands . . . are just some approaches that require a serious investment of public resources. . . . This sort of community development and reinvesting may appear politically infeasible. Yet for those who seek meaningful employment, . . . reduced utility bills, or more comfortable homes, such an agenda could prove not only salient but highly attractive.[35]

Public hesitation due to the difficulty of ending global warming can be overcome, especially if the public becomes convinced that they will share in the benefits.

Primary Source Quotes*

What Are the Solutions for Global Warming?

"[Global warming] is a very serious issue, and certainly, it is one of the major drivers of energy policies worldwide. We think that the issue is so serious that everything has to be done."

—Claude Mandil, "Instability Drops, Oil Follows Suit: Mandil," Australian Broadcasting Corporation Online, September 17, 2006.

Mandil is the executive director of the International Energy Agency.

"The entry into force of the Kyoto Protocol is a vital step toward this goal [of combating climate change]. . . . The Kyoto Protocol will provide the basis on which the community of states must endeavour to further develop climate protection in the years to come."

—Gerhard Schröder, "Message of Greeting for the British-German Climate Change Conference," Climate Change: Meeting the Challenge Together, Berlin, November 3, 2004.

Schröder is the former chancellor of Germany.

* Editor's Note: While the definition of a primary source can be narrowly or broadly defined, for the purposes of Compact Research, a primary source consists of: 1) results of original research presented by an organization or researcher; 2) eyewitness accounts of events, personal experience, or work experience; 3) first-person editorials offering pundits' opinions; 4) government officials presenting political plans and/or policies; 5) representatives of organizations presenting testimony or policy.

“As to some countries, [the Kyoto Protocol] did not require very much, as to some others, it required more than they could achieve. Some of its mechanisms . . . are proving to present great challenges.”

—James Connaughton, press briefing, May 18, 2005.

Connaughton is the senior environmental policy adviser to the Bush White House.

“New power plants . . . will be capable of capturing carbon dioxide for long-term storage. . . . [These] advances could also help to reduce emissions in countries like China and India.”

—Americans for Balanced Energy Choices, “Responding to Climate Change.” www.learnaboutcoal.org.

Americans for Balanced Energy Choices is a nonprofit organization founded by the coal-based electricity industry. ABEC seeks to educate the public about the benefits of coal-based power.

“The stakes [of reducing climate change] are high, but so are the potential rewards, including new investments in efficient technology . . . [and] real progress in reducing greenhouse gas emissions that will help address climate change.”

—Tony King and Steve Howard, “The Bottom Line on Global Warming,” *San Francisco Chronicle*, July 31, 2006.

King is the CEO of the Pacific Gas and Electric Company. Howard is the CEO of the nonprofit organization The Climate Group.

“[To] protect the climate, the United States must start cutting global warming pollution now and reduce emissions steadily over the coming decades.”

—Frances G. Beinecke, “New Global Warming Bill Signals Momentum Growing for Effective Reductions,” press release, Natural Resources Defense Council, January 12, 2007.

Beinecke is the president of the environmental advocacy group Natural Resources Defense Council.

"A few years back I proposed a regional greenhouse gas initiative, and that is now the law. We now have eight states, seven that are members, and an eighth, Maryland, which is going to be joining next year."

—George Pataki, "Governor's Remarks: Governors Schwarzenegger, Pataki Discussing Efforts to Curb Global Greenhouse Gas Emissions in New York City," Office of the Governor of the State of California, October 16, 2006. http://gov.ca.gov.

Pataki is the governor of the state of New York.

"[It] was certainly no surprise to us [at Waste Management] to see California to be the first state to impose a reduction goal on greenhouse gas emissions. . . . We really believe there's a world of opportunities for us."

—Larry O'Donnell, "Governor Arnold Schwarzenegger Signing Landmark Legislation to Reduce Greenhouse Gas Emissions," speech, Office of the Governor of the State of California, September 27, 2006. http://gov.ca.gov.

O'Donnell is the CEO of Waste Management, a trash and waste service company.

"AIG recognizes the scientific consensus that climate change is a reality and is likely in large part the result of human activities that have led to increasing concentrations of greenhouse gases in the earth's atmosphere."

—American International Group, "AIG's Policy and Programs on Environment and Climate Change," May 2006.

AIG is a leading insurer of businesses and individuals.

"I also think we need to work much harder to find ways to implement the vast range of low-carbon technologies that have already been developed. . . . [This] can be done, and often at a much lower cost than we realise."

—Tony Blair, "Davos Speech," World Economic Forum, January 26, 2005.

Blair is the former prime minister of Britain.

"There are economic consequences [to tackling global warming]. The problem is [the Bush administration has] been too quick to assume that such consequences will be bad for the economy. . . . Reducing greenhouse gases can be a serious moneymaker."

—Hal Harvey, "Combating Global Warming Makes Economic Sense," *San Francisco Chronicle*, February 21, 2006.

Harvey is the director of the Environment Program at the Hewlett Foundation.

"In the UK, we believe we can effectively tackle the challenge of climate change without large economic cost. We have already shown that action on climate change does not automatically lead to reduced wealth."

—Margaret Beckett, speech, Climate Change: Meeting the Challenge Together, Berlin, November 3, 2004.

Beckett is the secretary of state for environment, food and rural affairs in the United Kingdom.

What Are the Solutions for Global Warming?

- The United Nations reported that of the 41 countries under its watch, 34 had ***increased*** emissions during 2000–2004. These include countries that had signed the Kyoto Protocol.

- By 2050, **55 percent** of total greenhouse gas emissions could be exclusively from developing nations, according to the International Energy Agency.

- If America signed Kyoto, it would be required to limit its annual carbon emissions to **7 percent** below 1990 levels. This requirement does not set the country apart from its peers. The European Union is supposed to produce carbon at **8 percent** below 1990 levels, and Japan **6 percent.**

- In order to avoid a doubling of preindustrial levels of carbon dioxide, for instance, seven of Robert Socolow's stabilization wedges would be needed, each one mitigating **one gigaton** of carbon per year.

- According to the Pew Center for Global Climate Change, the benefits of mitigating a doubling of greenhouse gas emissions would be between **$55 billion** and **$140 billion.**

- According to the *Stern Review*, stabilizing greenhouse gas emissions at the current level would cost about **half a trillion dollars.**

Controlling Emissions with a Stabilization Wedge

A stabilization wedge, shown here, can illustrate how much reduction will be needed to control fossil fuel emissions. The topmost line of the wedge is the amount emissions would continue to increase if left unchecked. The bottom line of the wedge shows emissions being kept flat and eventually declining.

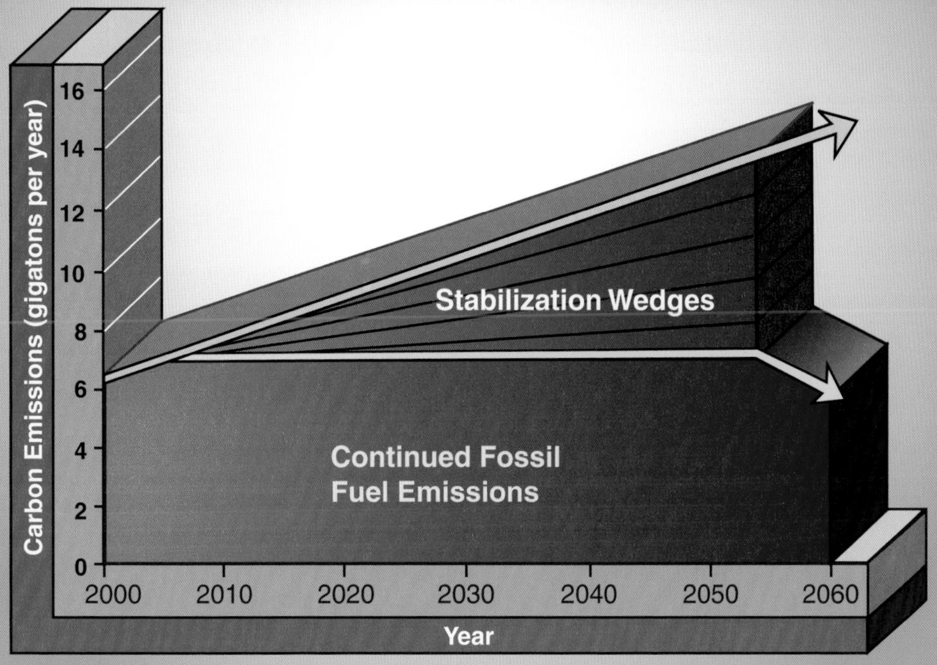

Source: S. Pacala and R. Socolow, *Science*, vol. 305, 2004.

- Renewable energy sources provide **2 percent** of the United States' energy needs, according to the Sierra Club.

- According to the American Wind Energy Association, less than 1 percent of the nation's energy is wind, compared to **20 percent** of Danish energy.

- According to the Union of Concerned Scientists, the cost of wind energy has decreased from 40 cents per kilowatt-hour in 1980 to between **3 and 6 cents** today.

- According to the U.S. Department of Energy, with about **9,733 megawatts (MW)** of installed capacity in 2002, biomass is the single largest source of nonhydro renewable electricity.

- In February 2002 President George W. Bush ordered a strategy to reduce the greenhouse gas emission intensity of the American economy by **18 percent** by 2012.

- California's Global Warming Solutions Act requires that the state cut greenhouse gas emissions **25 percent** to 1990 levels by 2020.

Wind Energy Capacity on the Rise

This graph shows that the global capacity to generate energy using wind has risen sharply from 1980 to 2005. Since 2001, world wind generation capacity has doubled due to reduced costs, supportive government policies, and concern about climate change.

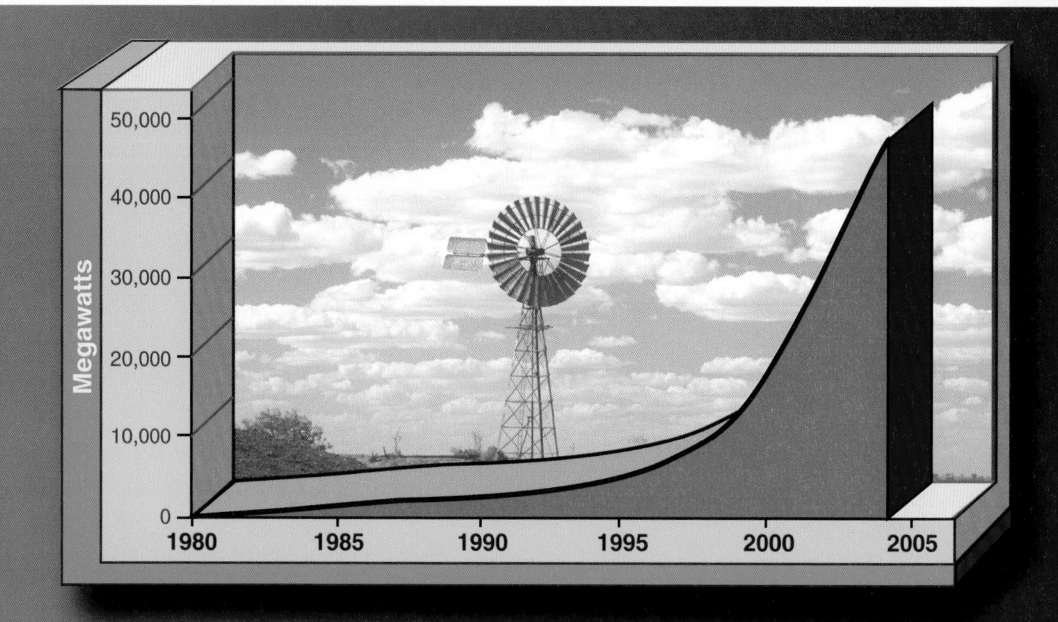

Source: Worldwatch Institute, "Energy: It's Blowin' in the Wind," October 19, 2005. www.worldwatch.org.

States Under Regional Greenhouse Gas Initiatives

This map shows the states that are participating in voluntary regional initiatives that place limits on greenhouse gas emissions. Every state in the western half of the country and in New England, as well as New York is under at least one program.

*States with diagonal shading indicate two programs

West Coast Governors' Initiative

Southwest Climate Change Initiative

Powering Plants

Western Governors' Association

New England Governors and Eastern Canadian Premiers

Regional Greenhouse Gas Initiative

States with no initiatives

Source: Pew Center on Global Climate Change, *Climate Change 101: Overview*, 2006. www.pewcenter.org.

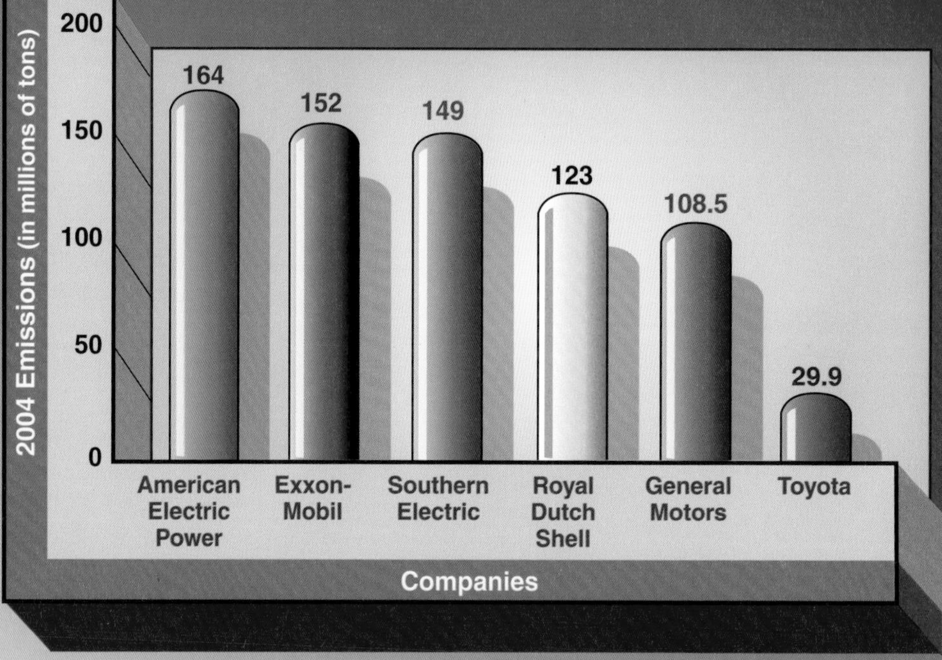

Source: *Business Week*, "Global Warming: Here Come the Lawyers," October 30, 2006. www.businessweek.com.

- United States greenhouse gas emissions increased **12 percent** between 1990 and 2001, according to the Pew Center on Global Climate Change.
- California's global warming act is the first set of **mandatory caps** on carbon emissions in the United States.
- According to the *San Francisco Chronicle*, China's greenhouse gas emissions were estimated at **97 percent** of the U.S. level in 2006, as compared to only **42 percent** of the U.S. level in 2001.

Key People and Advocacy Groups

Tony Blair: The former prime minister of the United Kingdom, Blair has made climate change a key issue during his time in office. In addition to imposing strict greenhouse gas emissions caps, Blair also commissioned the landmark *Stern Review on the Economics of Climate Change.*

Al Gore: Gore was vice president during the Clinton administration and has been an advocate for environmental protection since the late 1980s. During his time in office he pushed for passage of the Kyoto Protocol. In 2006 Gore wrote and starred in the anti–global warming documentary *An Inconvenient Truth*, which is credited with helping to popularize the climate change debate.

Jim Hansen: The director of NASA's Goddard Institute for Space Studies, Hansen is one of the world's leading experts on global warming and climate change. Hansen sparked controversy in 2006 when he accused the Bush administration of trying to stop him from speaking out about global warming.

James Inhofe: Inhofe is a Republican senator from Oklahoma and the former chair of the Senate Committee on the Environment and Public Works. He has been one of Congress's most vocal opponents of the scientific consensus on global warming. In a 2003 speech on the Senate floor, Inhofe called global warming "the greatest hoax ever perpetrated on the American people."

Intergovernmental Panel on Climate Change: The IPCC was established by the United Nations to study human-induced climate change and is led by Rajendra Pachauri. The panel's assessment reports are considered the authoritative source on the topic.

International Energy Agency: Led by Claude Mandil, the IEA is an energy policy adviser to countries around the world. With environmental protection as part of its mandate, the agency has made climate change a key priority.

Natural Resources Defense Council: The NRDC is one of the nation's leading nonprofit environmental groups. The 1.2-million-member organization has made ending global warming one of its top priorities and works with industry and the courts to achieve this goal.

Arnold Schwarzenegger: As the governor of California, Schwarzenegger has consistently worked in support of greenhouse gas emissions reductions. In 2006 Schwarzenegger signed the Global Warming Solutions Act, which imposed the first mandatory carbon dioxide caps in the United States and set a timeline for a significant reduction in California's greenhouse gas emissions.

Fred Singer: Singer is a physicist and leading skeptic on human-induced global warming. He is head of the Science and Environmental Policy Project, a group that disputes the scientific consensus on anthropogenic climate change, favoring an explanation focused on natural environmental changes.

Robert Socolow: A professor of engineering at Princeton University, Socolow, along with Princeton professor Stephen Pacala, invented the concept of stabilization wedges, a system of graphing emissions-reduction techniques.

Nicholas Stern: Stern is a British economist and author of the landmark 2006 *Stern Review*, which was commissioned by the British government and which analyzes the effects of climate change from a social and economic viewpoint.

Woods Hole Oceanographic Institution: A nonprofit oceanic science research corporation located in Woods Hole, Massachusetts, WHOI research has provided important data on the effects of global warming on the world's oceans and marine life.

Chronology

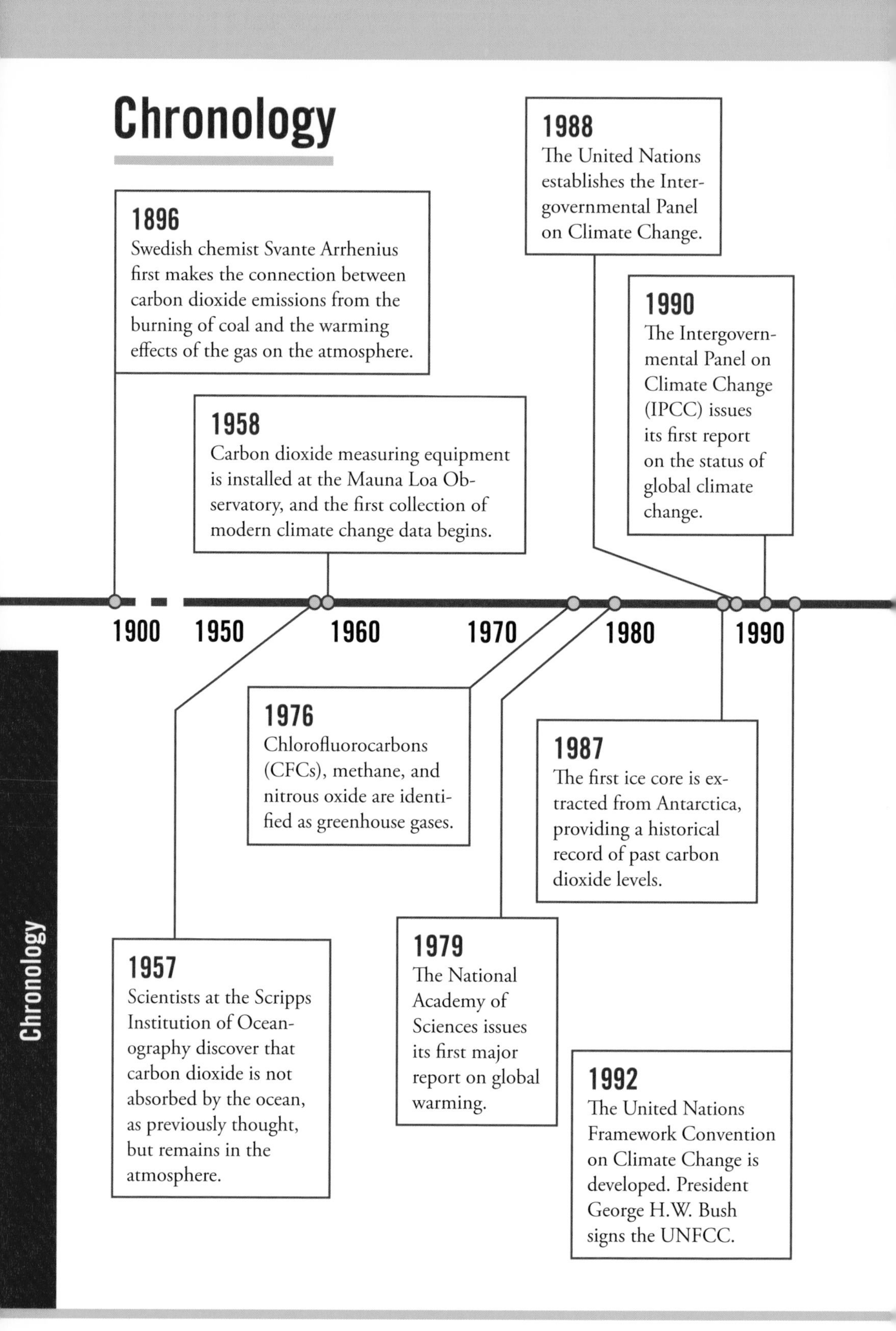

1896
Swedish chemist Svante Arrhenius first makes the connection between carbon dioxide emissions from the burning of coal and the warming effects of the gas on the atmosphere.
1958
Carbon dioxide measuring equipment is installed at the Mauna Loa Observatory, and the first collection of modern climate change data begins.
1988
The United Nations establishes the Intergovernmental Panel on Climate Change.
1990
The Intergovernmental Panel on Climate Change (IPCC) issues its first report on the status of global climate change.
1900
1950
1960
1970
1980
1990
1976
Chlorofluorocarbons (CFCs), methane, and nitrous oxide are identified as greenhouse gases.
1987
The first ice core is extracted from Antarctica, providing a historical record of past carbon dioxide levels.
1957
Scientists at the Scripps Institution of Oceanography discover that carbon dioxide is not absorbed by the ocean, as previously thought, but remains in the atmosphere.
1979
The National Academy of Sciences issues its first major report on global warming.
1992
The United Nations Framework Convention on Climate Change is developed. President George H.W. Bush signs the UNFCC.
Chronology

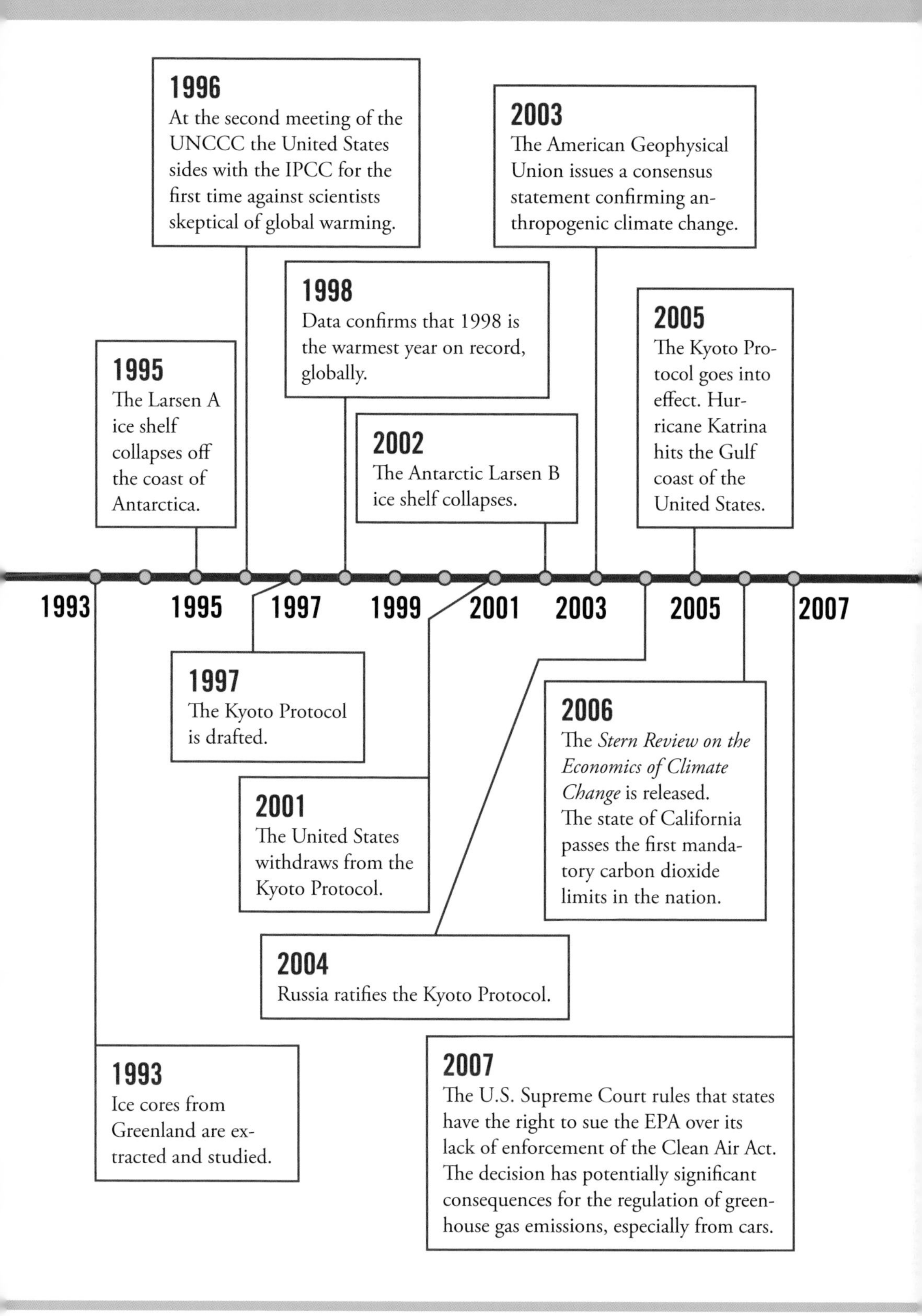
1996
At the second meeting of the UNCCC the United States sides with the IPCC for the first time against scientists skeptical of global warming.
2003
The American Geophysical Union issues a consensus statement confirming anthropogenic climate change.
1998
Data confirms that 1998 is the warmest year on record, globally.
2005
The Kyoto Protocol goes into effect. Hurricane Katrina hits the Gulf coast of the United States.
1995
The Larsen A ice shelf collapses off the coast of Antarctica.
2002
The Antarctic Larsen B ice shelf collapses.
1993
1995
1997
1999
2001
2003
2005
2007
1997
The Kyoto Protocol is drafted.
2006
The Stern Review on the Economics of Climate Change is released. The state of California passes the first mandatory carbon dioxide limits in the nation.
2001
The United States withdraws from the Kyoto Protocol.
2004
Russia ratifies the Kyoto Protocol.
1993
Ice cores from Greenland are extracted and studied.
2007
The U.S. Supreme Court rules that states have the right to sue the EPA over its lack of enforcement of the Clean Air Act. The decision has potentially significant consequences for the regulation of greenhouse gas emissions, especially from cars.

Related Organizations

American Geophysical Union (AGU)

2000 Florida Ave. NW

Washington, DC 20009

phone: (202) 462-6900

fax: (202) 328-0566

Web site: www.agu.org

The AGU is a nonprofit organization of geophysicists dedicated to the dissemination of information about the field. In 2003 the AGU issued a consensus statement confirming the presence of anthropogenic climate change. The AGU publishes the weekly journal *Eos*.

Center for the Study of Carbon Dioxide and Global Change

PO Box 25697

Tempe, AZ 85285

phone: (480) 966-3719

fax: (480) 664-4923

e-mail: staff@co2science.org

Web site: www.co2science.org

This nonprofit organization seeks to spread factual information and sound science about the consequences of carbon dioxide in the atmosphere. The group's stated goal is to separate fantasy from reality in the climate change debate. The center publishes the weekly online magazine *CO_2 Science.*

Environmental Protection Agency (EPA)

1200 Pennsylvania Ave. NW

Washington, DC 20460

phone: (202) 272-0167

Web site: www.epa.gov

The EPA is a federal agency charged with protecting human health and the environment. The EPA develops environmental regulations, performs research, and educates the public about key issues related to the environment.

National Center for Atmospheric Research (NCAR)

PO Box 3000

Boulder, CO 80307

phone: (303) 497-1000

Web site: www.ucar.edu

NCAR is a division of the nonprofit University Corporation for Atmospheric Research. The center provides research, facilities, and services for climate and atmospheric scientists.

National Oceanic and Atmospheric Association (NOAA)

14th St. & Constitution Ave. NW, Room 6217

Washington, DC 20230

phone: (202) 482-6090

fax: (202) 482-3154

e-mail: answers@noaa.gov

Web site: www.noaa.gov

NOAA is a division of the U.S. Department of Commerce. The agency focuses on researching the conditions of the oceans and atmosphere, as well as supplying information about the weather and atmospheric conditions to the public.

Pew Center on Global Climate Change

2101 Wilson Blvd., Suite 550

Arlington, VA 22201

phone: (703) 516-4146

fax: (703) 841-1422

Web site: www.pewclimate.org.

The Pew Center is a nonprofit organization that seeks to provide objective research and analysis on the climate change issue. The center works with industry leaders, governmental bodies, and scientists to craft innovative climate change solutions.

Science and Environmental Policy Project (SEPP)

1600 S. Eads St., Suite 712-S

Arlington, VA 22202

e-mail: comments@sepp.org

Web site: www.sepp.org

SEPP is a nonprofit organization founded by physicist Fred Singer to encourage scientifically sound, cost-effective decisions on health and the environment. SEPP promotes a theory of climate change based on natural causes.

Union of Concerned Scientists (UCS)

2 Brattle Sq.

Cambridge, MA 02238

phone: (617) 547-5552

fax: (617) 864-9405

Web site: www.ucsusa.org

The UCS is a nonprofit advocacy group composed of citizens and scientists who work to promote a healthy environment and a safe world. In addition to climate change, the union has programs focused on clean air, clean water, and invasive species, among others.

World Meteorological Organization (WMO)

7bis, Ave. de la Paix, Case Postale No. 2300, CH-1211

Geneva 2, Switzerland

phone: +41 22 730 81 11

fax: +41 22 730 81 81

e-mail: wmo@wmo.int

Web site: www.wmo.int

The WMO is a United Nations agency that functions as the UN's voice on issues of atmosphere, climate, and weather. The WMO is one of two agencies that were instrumental in establishing the Intergovernmental Panel on Climate Change.

Worldwatch Institute

1776 Massachusetts Ave. NW

Washington, DC 20036

phone: (202) 452-1999

fax: (202) 296-7365

e-mail: worldwatch@worldwatch.org

Web site: www.worldwatch.org

The Worldwatch Institute is an independent, nonprofit research organization that studies a variety of environmental issues, including climate change. The institute publishes the bimonthly *World Watch* magazine as well as an annual report, *Vital Signs.*

For Further Research

Books

Richard B. Alley, *The Two-Mile Time Machine: Ice Cores, Abrupt Climate Change, and Our Future*. Princeton, NJ: Princeton University Press, 2002.

David Archer, *Global Warming: Understanding the Forecast*. Ames, IA: Blackwell, 2006.

John Cox, *Climate Crash: Abrupt Climate Change and What It Means for Our Future*. Washington, DC: Joseph Henry, 2005.

Kirsten Dow and Thomas Downing, *The Atlas of Climate Change: Mapping the World's Greatest Challenge*. Berkeley and Los Angeles: University of California Press, 2006.

Al Gore, *An Inconvenient Truth: The Planetary Emergency of Global Warming and What We Can Do About It*. Emmaus, PA: Rodale, 2006.

John Houghton, *Global Warming: The Complete Briefing*. Cambridge: Cambridge University Press, 2004.

Thomas E. Lovejoy, ed., *Climate Change and Biodiversity*. New Haven, CT: Yale University Press, 2006.

Mark Lynas, *High Tide: The Truth About Our Climate Crisis*. New York: Picador, 2004.

Chris Spence, *Global Warming: Personal Solutions for a Healthy Planet*. New York: Palgrave Macmillan, 2005.

David Victor, *Climate Change: Debating America's Policy Options*. New York: Council on Foreign Relations, 2004.

Periodicals

Michael Arndt, "How Dupont Grew Greener," *BusinessWeek Online*, October 16, 2006.

John Carey, "Global Warming Heats Up Capitol Hill," *BusinessWeek*, September 18, 2006.

———, "Ocean Currents: Flipping the 'Off' Switch," *BusinessWeek*, October 18, 2006.

John Cassidy, "High Costs," *New Yorker*, November 13, 2006.

Gwen Glazer, "Global Warming," *National Journal*, September 2, 2006.

Marc Gunther, "The Governator Takes Aim at Global Warming," *Fortune*, September 18, 2006.

Mark Hertsgaard, "CA Leads on Climate," *Nation*, October 2, 2006.

Michelle Higgins, "Raising the Ante on Eco-Tourism," *New York Times*, December 10, 2006.

Frank Keppler and Thomas Rockmann, "Methane, Plants and Climate Change," *Scientific American*, February 2007.

Elizabeth Kolbert, "Hot and Cold," *New Yorker*, December 11, 2006.

Marianne Lavelle, "The Market to Clear the Air," *U.S. News & World Report*, December 18, 2006.

Mark Lynas, "Australia's Climate Change Shame," *New Scientist*, November 20, 2006.

Jessica Marshall, "Glaciers Heading for Point of No Return," *New Scientist*, August 19, 2006.

Fred Pearce, "'One Degree and We're Done For,'" *New Scientist*, September 30, 2006.

Bret Schulte, "A Storm over Warming," *U.S. News & World Report*, September 4, 2006.

Karen Sue Smith, "All Heated Up," *America*, December 11, 2006.

Gary Stoller, "Concern Grows over Pollution from Jets," *USA Today*, December 19, 2006.

Doug Struck, "Climate Change Drives Disease to New Territory," *Washington Post*, May 5, 2006.

Stuart Taylor Jr., "Global Warming: Time for a Court Order," *National Journal*, December 2, 2006.

John Young, "Black Water Rising," *World Watch*, September/October 2006.

Internet Sources

Catherine Brahic, "Carbon Emissions Rising Faster than Ever," *New Scientist Environment*, November 10, 2006. http://environment.newscientist.com.

Climate Ark, "Climate Change Overview." www.climateark.org.

David Easterling, "Global Warming: Frequently Asked Questions," National Oceanic and Atmospheric Administration, February 3, 2006. www.ncdc.noaa.gov.

Mark Henderson, "New Proof That Man Has Caused Global Warming," *Times Online*, February 18, 2005. www.timesonline.co.uk.

John Collins Rudolf, "The Warming of Greenland," *NYTimes.com*, January 16, 2007. www.nytimes.com.

Source Notes

Overview

1. David A. King, "Climate Change Science: Adapt, Mitigate, or Ignore?" *Science*, January 9, 2004.
2. John P. Holdren and Alan I. Leshner, "Time to Get Serious About Climate Change," *San Francisco Chronicle*, July 30, 2006.
3. Mark Lynas, "Australia's Climate Change Shame," *New Statesman*, November 20, 2006.
4. *Bulletin of the American Meteorological Society*, "Climate Change Research: Issues for the Atmospheric and Related Sciences," February 9, 2003.
5. Pallab Ghosh, "Caution Urged on Climate 'Risks,'" BBC News, March 17, 2007. www.news.bbc.co.uk.
6. Joke Waller-Hunter, "Statement by Executive Secretary of the United Nations Framework on Climate Change at the Twenty-Second Session of the IPCCC," November 2004.

What Is Global Warming and Climate Change?

7. *Bulletin of the American Meteorological Association*, "Climate Change Research."

What Are the Consequences of Global Warming?

8. Don Walsh, "Losing Its Cool," *U.S. Naval Institute Proceedings*, January 2007.
9. Don Walsh, "Losing Its Cool."
10. John Young, "Black Water Rising," *World Watch*, September/October 2006.
11. National Center for Atmospheric Research, "Arctic, Antarctic Melting May Raise Sea Levels Faster than Expected," March 23, 2006.
12. Johannes Oerlemans et al., "Ice Sheets and Sea Level," *Science*, August 25, 2006.
13. World Bank, "Climate Changes and Impact on Coastal Countries," February 12, 2007. http://econ.worldbank.org.
14. *New Scientist*, "Arctic Ocean Continues to Freshen Up," September 5, 2006.
15. Richard A. Kerr, "False Alarm: Atlantic Conveyor Belt Hasn't Slowed Down After All," *Science*, November 17, 2006.
16. William Curry, "Common Misconceptions About Abrupt Climate Change," Woods Hole Oceanographic Institution. www.whoi.edu.
17. Quoted in Fred Pearce, "'One Degree and We're Done For,'" *New Scientist*, September 30, 2006.
18. Quoted in S. Perkins, "On the Rise," *Science News*, September 9, 2006.
19. Quoted in D.T. Max, "Making Global Warming Work for You," *New York Times*, December 11, 2005.
20. Dennis T. Avery and H. Sterling Burnett, "Global Warming Famine—or Feast?" National Center for Policy Analysis, May 19, 2005.

What Are the Consequences of Global Warming?

21. L.F. Khilyuk and G.V. Chilingar, "On Global Forces of Nature Driving the Earth's Climate. Are Humans Involved?" *Environmental Geology*, August 2006.

22. Eugene Linden, *Winds of Change*. New York: Simon & Schuster, 2006.
23. Quoted in Chris Mooney, "Climate Challenge," *American Prospect*, April 4, 2005.
24. James Inhofe, "Climate Change Update," Senate floor speech, January 4, 2005. www.inhofe.senate.gov.
25. Thomas J. Crowley, "Are Claims of Global Warming Being Suppressed?" *Eos*, February 7, 2006.
26. American Association for the Advancement of Science, "AAAS Board Statement on Climate Change," December 9, 2006.
27. Myron Ebell, "Love Global Warming," *Forbes*, December 25, 2006.
28. Tom Bethell, "The False Alert of Global Warming," *American Spectator*, May 2005.
29. Robert Samuelson, "The Worst of Both Worlds?" *Newsweek*, November 13, 2006.
30. Samuelson, "Worest of Both Worlds."

What Are the Solutions for Global Warming?

31. Holdren and Leshner, "Time to Get Serious About Climate Change."
32. Marc Gunther, "Insurance Companies Take On Global Warming," *CNN Money.com*, August 24, 2006. http://money.cnn.com.
33. John Carey, "Global Warming Heats Up Capitol Hill," *Business Week*, September 18, 2006.
34. Dianne Feinstein, speech, Commonwealth Club, San Francisco, California, August 24, 2006.
35. John M. Meyer, "Another Inconvenient Truth," *Dissent*, Fall 2006.

List of Illustrations

Index